The DNA Delusion

By Dr. Stephen T. Blume'

Published by:
Create Space Publishing, 2018

Copyright © 2017 Dr. Stephen T. Blume
All rights reserved.
ISBN: 1976077729
ISBN-13: 9781976077722

Dedication:

This book is dedicated to Richard Dawkins, the world's leading evo-illusionist. Dawkins has written many books on the subject of evolution. His most famous book, *The Blind Watchmaker*, has certainly taken the place of Darwin's *On the Origin of Species by Natural Selection* as the Bible of evolution. He has done a great job of fooling his readers and fans into believing the most incredible fables. He has his own organization, *The Richard Dawkins Foundation* that will survive long after Dawkins is gone. His followers treat him with reverence, much like the pope is treated with great reverence by his followers. Also included in my dedication are all of the evo-illusionists who have made my scientific explorations and investigations so much fun and so very fascinating. They fooled me when I was in college to such a degree that I became an avid fan of evolution, what I call an *evolutionaut*, for many decades.

My first exposure to evolution was in my freshman biology and anthropology classes at the University of Southern California. I had such an excited and fascinated feeling. Evolution made so much sense. The immense mystery that I had puzzled over for so long was gone forever. Well, I *believed* it was gone anyway. For years I argued for evolution like any dedicated fan would. Everyone who was not an evolutionaut was wrong. I *knew* how everything formed. Those that weren't believers and supporters of evolution didn't know; plain and simple. Evolution gave me a kind of high, a superiority, an excitement that kept growing with each new book on the subject that I read, and with each new documentary that I watched. I was in a family of people who were similarly enlightened; who understood oh so much. I can thank evo-illusionists like Richard Dawkins for my first round of great excitement and enlightenment. I have always loved science and the puzzles it presented. When one of those immense puzzles was solved for me, it was truly a memorable moment; it was pure fun.

When my son entered medical school in Chicago, I visited him there many times. My trips to Chicago gave me the opportunity to visit Chicago's Field Museum of Natural History, which has the greatest fossil collection in the world. Of course I was thrilled to have the opportunity to actually see, close up, the fossil collection at the Field. I will never forget the feeling of excitement I had as I crossed the street to enter the museum for the first time; that eager emotion I had, knowing I was going to see new and fantastic fossil finds that would prove evolution to an even greater degree than I had ever experienced. I hadn't visited a really great natural history museum for several decades. Surely that time span would have turned up uncountable new fossil finds that would make evolution even more of a lock. I kind of felt like I was walking three feet off the ground. I couldn't wait to get inside.

After entering the museum with such high expectations, an awful kind of disappointment began setting in. As I went from exhibit to exhibit, room to room, I didn't see what I so eagerly expected to see. Where were the fossils that showed land animals evolving into whales? Fish that grew legs and walked on shore? Bird precursors that demonstrated the evolution of flight? Where were the gradual changes in species over millions of years? There were thousands of fossils that represented ancient plants and animals. But over vast timespans represented by the fossils, species all looked alike. There was no gradual change; not any. Where was it? In the next room? The next display? Another museum? There were 250 million year old frog fossils, but they were no different than modern frogs. There were trilobites over hundreds of millions of years without change. The same with fish, sea horses, insects... That euphoric feeling that I had walking to the museum just sank. The fossils that I was so excited about weren't there. The inarguable proof of evolution wasn't either. Where were they? I don't know if I was disappointed, or just puzzled. But what I saw really started me wondering; and questioning. Did I support a fantasy for all of these years?

I went home and began doing a great deal of research on the subject. Maybe evolution websites would show the changes in fossils that I must have certainly missed at the museum. But they didn't. What I found was excuses for the lack of change in the fossil record. For the first time I learned about *punctuated equilibrium*. All of the books I read and documentaries I watched on evolution had never mentioned punctuated equilibrium. I learned that changes happened so relatively fast that they didn't show up in the fossil record. After all, 100,000 years can be like a day when discussing fossils and geologic time. Punctuated equilibrium asked, what are the odds that a series of fossil changes would even show up since there are so few fossils compared to the populations of past species? Seemingly long periods time are, in reality, comparatively and geologically short. The changes we cannot see all somehow occurred in these geologically short times. There simply haven't been enough fossils found that demonstrate evolution. I learned that most species remained in equilibrium for long periods, and that the great changes were absent because they undoubtedly occurred during the relatively short "punctuated" periods. The odds were enormously against finding fossils from these comparatively short punctuated periods. But the problem with punctuated equilibrium was that the fossils on both sides of these punctuated time periods could be observed. And they were the same. Which means there was no change in those punctuated periods. So, for me, punctuated equilibrium made no sense. It was an excuse.

So I studied, did research, and wondered. Could evolution be no different than the tales I was taught as a child in church? Were they no different than the notion that Noah and his family collected all of the animals of the world on his wooden boat? Or that Adam and Eve were the first people on Earth? Did evolution have the same validity as these stories? The more I thought, the more the answer became a resounding "yes"? Evolution was no different than the ancient fables. We humans have no idea where all living species, body parts, and biological systems came from. The more I thought about evolution, the more I studied, the more skeptical I became. Until in a funny sort of way, I again felt enlightened; and excited. I had a whole new way of looking at the origins of all living things; and in particular, humans. I questioned, how could evolution produce vision? Was evolution capable of making an eye, with its autofocus lens, its auto stop-down iris, and its image-capturing retina? Could it pair up eyes so depth perception would result? Did evolution even know what depth perception was? Could it make an optic nerve, and a visual cortex in the brain so chemical signals in the retina could be transposed into the perception of a color image? Could evolution produce perception? When I researched evolution websites, I found nice diagrams showing how eyes evolved. But they never discussed the optic nerve and visual cortex. Why are these never included in the diagrams? The diagrams looked like instructions for the assembly of a digital camera. "First you insert this part. Next you…" The eye in particular was actually such a fun thought. No matter how hard I tried, I couldn't come up with any sort of fable, no matter how ridiculous, about how eyes and vision might have resulted from gradual changes. Each time I thought out how, in step-by-step fashion, any physiological system might have evolved, I became stuck. There weren't even any imaginary stories I could come up with that might produce any biological system through the slow random changes of evolution.

My new enlightenment followed me around wherever I went. One day while walking outside, I walked right into a spider web. I immediately began considering its evolution. Did the web-making device evolve first, and then the spider began forming a web? It had to happen that way. Why would the web-making device form in spiders at all until they could actually form a web? Was there a use for partially formed web-making apparatus? Did spiders immediately form full webs as soon as their apparatus was functional? How did that happen? How were their brains programmed to make webs? Was the first web just a string from one point to another? Did spider's web-making program also evolve to become very complex? I kind of made a game out of evolution. By observing various physiological systems in nature, I tried to form some sort of fable for how they came to be through slow random changes. How did cocoons evolve? How did butterflies evolve their wing designs, beautiful artwork, and crazy flight patterns? I completely struck out. There were no plausible fables I could even make up for the evolution of any biological system at all. Of course I searched evolution websites for answers, but there were none. Actually, it was amusing trying to figure out stories for evolution and how it could have formed every biological system I ran into. The most amusing thing is I couldn't come up with any stories.

I felt enlightened in a whole new and exciting way. The fact that evolution was an obvious fable meant we humans have no conceivable story for how all of living nature came to be. And so I had my second very euphoric moment; a second time in my life that evolution got my head spinning with thoughts about origins; about how we humans came into being. The first was when I realized evolution produced all living things. The second was many years later, when I realized it didn't, and that millions of people were just as gullible as I was because they believed the fable and Richard Dawkins. I was fortunate to have two periods in my life of such excitement. They were both so much fun, so interesting and thought provoking. And all of this was because Richard Dawkins and evo-illusionists like him seriously supported, wrote about, and promoted such an absurd fable; because Dawkins and his fellow evo-illusionists have been able to fool millions of people, including me, into believing. When I became a true believer in evolution, I had my first very excited and thought provoking moment. A moment that actually lasted a long time. Then, when I realized evolution was a fable, and what I had believed for so many years was a fraud, I had my second. Anyone who has held a belief for many years, and realized what they believed was completely wrong, will understand the mind games beliefs have played on me.

Since the time I realized that I had been so fooled by Richard Dawkins and his fellow evo-illusionists, I have written a blog, *www.evoillusion.org*, made many YouTube videos under the pseudonym stevebee92653, and wrote three books on the subject including this one. These are all things I love doing. So Dawkins and friends have given me more fun than anyone deserves, more excitement, happiness, and a real *raison d'etre*; a purpose. They changed my life and thinking in a remarkable way. In pondering who I should dedicate this book to, I can't come up with a more deserving person than Richard Dawkins and his fellow evo-illusionists. So here's to you Richard. This book is officially dedicated to you; and your fellow evo-illusionists. Please feel free to let them know. Thanks for the fun! I truly hope you will read this book and let me know where you think I am right or wrong. But I know the chances of that are about $1:10^{190}$.

Prologue:

If you are not familiar with what DNA really is, the letters represent the molecule *deoxyribonucleic acid*. So now you see why DNA is *deoxyribonucleic acid's* commonly utilized nickname. We are used to hearing the term "DNA" on police oriented television dramas, and in movies. DNA is very specific to every individual, and it can be used to identify the bad guys in criminal investigations. It is far more accurate than are fingerprints. DNA carries critical coding that makes life itself possible. DNA is tightly wrapped in the nucleus of the cells of every living multicellular organism. Few people know what it really does, so in a way it's still a mystery to most people. Few people know that DNA makes up our genes, and that genes make up our chromosomes; that they are all part of a very important whole.

Skin cells are so small that 10,000 of them will fit on the head of a pin. But if you could unravel and straighten out the DNA of one basic skin cell, it would be over six feet long. In all multicellular species, DNA is wrapped perfectly and unbelievably tightly around tiny spools called *histones* so it can fit inside of the nucleus, the center part of those tiny cells. A skin cell's *nucleus* is about half of the diameter of the skin cell. To better understand DNA, let's expand this whole thing up into entities that are sizes we are more used to. If an ordinary skin cell could be expanded to the diameter of a basketball, 9.55 inches, the DNA inside of that basketball-cell would be 9,188 miles long. It's just unimaginable to think about. That's more than 1/3 of the way around the Earth. The nucleus of our imaginary basketball-cell would be a sphere the size of an orange. Can you imagine trying to stuff thread that is 9,188 miles long inside of a 4-inch sphere like nature does on a ho-hum routine basis every microsecond of every day? This incredible system carries coding for the most important biochemicals in our bodies, and the bodies of all living things. As I said, DNA makes up our genes, which in turn make up our chromosomes. Without DNA there would be no life. If alien life has ever formed, it would most certainly have to be based on DNA and the products it codes for.

In 1953, two scientists, James Watson and Francis Crick successfully solved one of science's greatest puzzles. They did the earth-shaking task of discovering and determining the molecular structure of DNA. It was immediately thought and pronounced that what they did was solve the mystery of what holds the blueprints, the plans for the entire human body. They had found the "Secret to Life!" Evolution science went into a celebratory frenzy. Finally the basis for evolution was found. Evolution was proved. Since DNA carries the plans for the entire human body, and the body of all species, tiny errors in the copying of DNA when it is passed on from parent to offspring causes small body changes which leads to the formation of all species and their body parts.

So evolution science swallowed up and engulfed DNA as its own. The discovery of the DNA molecule and its structure produced an entirely new and modern version of evolution. In the middle of the 19th century when Darwin made his theory, he had no idea what caused the changes that he observed in birds in the Galapagos Islands. So while his theory became more and more accepted, there was still an immense and puzzling hole to fill. What was the *modus operandi* for those changes in Darwin's finches? Watson and Crick filled that immense hole, and modern evolution was born.

DNA became the *god molecule* for so many sciences; the *savior*. DNA explained what directs the biochemistry of all cells. Genes that are made up of DNA direct the formation of all human infants from a single fertilized egg. DNA in the form of genes holds the blueprints for the entire human body, so that mystery was solved. DNA acts like a brain, and directs the entire goings on inside of cells. So many scientific puzzles were deemed solved with the solving of the DNA puzzle. Pretty much every educated person now believes that "it's all in the genes". All of our abilities, our appearance and the reason we look like our parents, our personalities, our traits and characteristics, the design of our body parts such as our skeletons, visual systems, auditory systems... everything was credited to our genes. I personally have asked many people the question about what directs the formation of the human body. The answers have been 100% identical. "Oh, it's our genes." Children in school are taught that. It's repeated many times all along the way. Virtually every government science organization is sure "it's all in the genes."

This book takes a very scientific and objective look at that theory. Are our genes all they are cracked up to be? Can they do all of what we have been taught they can? Is the mystery of what directs the maturing of an infant from a single cell really directed by genes? Are we alive and well because of the tasks performed by our DNA? Are all of our inherited characteristics due to our genes? Do we look like our parents because of our genes? Or, is there some other entity involved. Do micro-changes in our DNA during procreation result in the vast number of plant and animal species on Earth, and all of their body parts and biochemical systems? Can evolution truly claim changes in DNA over generations and eons are its driving force? Is DNA truly the answer to so many mysteries of life? I wrote this book to answer these very demanding and fascinating questions. I answer them in a purely scientific and objective basis. But I do answer them. I hope this book will answer them for you, and possibly change your thinking about our genes, our DNA, and what they can and cannot do.

Table of Contents

Dedication		3
Prologue		7
Table of Contents		9
Chapter 1	It's All in the Genes?	11
Chapter 2	The Search for DNA Begins	18
Chapter 3	Finally a Breakthrough	26
Chapter 2	OK Steve, Why Doesn't DNA Hold the Plans For Our Bodies?	38
Chapter 5	DNA and Friends: The First True Computer Hard Drive	46
Chapter 6	DNA and How it Works	59
Chapter 7	A Sheep and a Cell Bid Adieu to the God Molecule	71
Chapter 8	Are We Just a Bunch of Mistakes?	79
Chapter 9	The Inner and Outer Domains Further Destroy The Illusion of DNA	87
Chapter 10	So Then What Makes Babies?	91
Chapter 11	Is The Blind Watchmaker Really Blind?	107
Chapter 12	The Abio-alchemist Stirred the Pot and...	119
Chapter 13	Spontaneous Generation Just Won't Go Away	130
Index		143
References		146

Chapter 1

It's All in the Genes?

We've discovered the secret of life.—Francis Crick

This book, and my two previous books, *Evo-illusion*, and *Evo-illusion of Man*, are discussions about the science of origins of the universe and living species, and the validity of evolution. They take a purely scientific, objective, and critical look at what modern science has to say about the source of all living species, including humans. There is no religious bent to any of my books whatsoever. They are not part of an evolution versus creationism debate. My writings only are concerned with the validity of evolution and the pertinent sciences that are utilized to surround and support it. My writings don't promote the interests or cause of any religion or any ancient notions about how all living creatures originated. The science in my books is pretty fundamental, and can be understood by most people who've had very basic high school biology. The key to, and the entire foundation for evolution, is the accidental changes that occur in the genetic systems of all living organisms from their beginnings; from basic one-celled species, and before, to the multi-celled species of today. Obviously, it would be a good idea to determine if the basis for, in fact the whole underpinning of evolution, is valid. To really understand evolution, you have to have a basic knowledge regarding the genetic system of all living organisms. The backbone of that system is DNA. If you can understand that a 10-kilobyte computer cannot do a job that requires 1,000 terabytes, you will understand what I have to say about DNA.

The story about the unraveling of the structure of DNA is a fantastic one. Super scientific sleuths exposed the mystery of DNA by gradually collecting hints that were oh so slowly doled out by an amazing sub-microscopic world that is thousands of times busier than Manhattan during peak traffic time. Once the collection of clues, which were uncovered by numerous researchers over many decades, became sufficient, two of the most unlikely scientific super sleuths put those clues together and solved the seemingly unsolvable puzzle of the molecular structure of DNA. DNA and the puzzle it presented exists in a vast world that is so miniaturized that few people on Earth, other than knowing DNA exists, have any idea what it actually is or does. The best microscopes in the world can't come close to imaging a DNA molecule. The solution to the DNA puzzle came without the two sleuths who solved it ever actually seeing any parts of DNA itself. Imagine doing a jigsaw puzzle without ever seeing or touching any of the puzzle parts.

DNA does fantastic things that we humans cannot observe. These fantastic things are going on by the trillions in every living cell in our bodies every second of the day. DNA, and the puzzle it presented, is at least as fascinating as the moon landing. The rewards and benefits that resulted from discovering the structure of

The DNA Delusion

the DNA molecule far outshine the benefits and ramifications of the moon landing.

But the moon is always there for us to see. What could be more beautiful than a huge full moon just above the horizon? The imagination of mankind that the moon has stirred up, the excitement created by the dream of someday going to the moon, and the plethora of stories and songs written about the dreamlike countenance of the moon, made the landing of man on the moon one of the greatest moments in human history. Those that experienced

Fig. 1-1

the first lunar landing on television will never forget where they were the moment Neil Armstrong said to the world, "Houston, Tranquility Base here. The Eagle has landed!" It kind of gives me chills to think about. I was in the Navy on a frozen island 1200 miles west of Anchorage, Alaska when I heard the landing described on the radio. Because of my location, I wasn't able to see it on television, but I was just as thrilled as if I did. I was just astounded that we could actually succeed at such a technical and perilous journey; that man was actually setting foot on the moon. The headlines about the lunar landing (Fig. 1-1) were huge and spread across page one of every newspaper on Earth.

On the other hand, the solving of the DNA puzzle happened quietly. I was eleven years old. Of course at that age, I had no idea this fantastic event occurred. No teachers announced to the class, "Children! The most exciting thing has just happened! We have uncovered the mystery of inheritance! We now know the molecular structure of DNA!" There were no students giving standing ovations because of this news. There were no excited dinner table chats about DNA with parents and kids. There were no tickertape parades. When it happened, few knew what DNA even was. So this

Fig. 1-2

phantasmagorical event took place in near silence. When man landed on the moon, billions of people were excitedly watching on television all over the world. It was as if the world was all together and focused on that one event for just that magic moment in time. When the news came out about the discovery of the solution to the structure of DNA, an event that would truly have ramifications for a huge number of the Earth's population, all was quiet. The news was printed on page 737 of *Nature* magazine. (Fig. 1-2) There were no newspaper headlines for DNA. Poor DNA was just too little, and too invisible.

In this book I intend to do a bit of catching up for DNA by hopefully creating excitement about its unusual molecular structure and the almost magical way it functions. The unimaginable things that DNA and the other molecules on its team do in every cell of our bodies on a routine basis are astonishing. So is the history of the unraveling of the structure of the DNA molecule. DNA is so mysterious and what it does is so incredible that, for certain, it rivals any manmade device, including the Saturn rocket and lunar lander. I fully realize that every person on Earth has looked at the moon in awe, and dreamed about its beauty and wonder. I also fully realize that no person on Earth, not one, can see DNA and *its* beauty and wonder. But if somehow a human cell could be expanded to the size of a basketball, and we could look inside to see what was going on, I guarantee you that DNA and its chemical partners would be far more mind-boggling than the moon. So I am going to commit the first two chapters of this book to the background of the discovery of DNA and how it's molecular structure and *modus operandi* were solved after over 100 years of trying. As I said, the story about the scientific sleuths who, so patiently, step by step, unraveled the mysteries of DNA, is a fantastic one.

DNA is a very important molecule to evolution science because it forms the foundation of evolution itself. It's evolution's *god molecule*. Evolution has elevated DNA to god status. DNA and changes in the DNA molecule during cell division is the great god creator of evolution. The solving of the puzzle of the structure and function DNA has turned evolution into a creationist religion. Richard Dawkins, the world's leading evo-illusionist, in his book *The God Delusion*, tries to prove there is no God. However Dawkins replaces God with his own version of a god: DNA. Evo-illusionists and most schools teach that our genes, which are made up of DNA, (1) hold the blueprint for the entire human body, and the bodies of all living organisms (2) hold the plans for all proteins of all species, and (3) are the control center of all cells. Every cell in our body has identical DNA code. Therefore each cell in our body holds the plans for the entire human body; so they say. Dawkins, in his bestselling book, *The Selfish Gene* says:

> There are about a thousand million million cells making up an average human body, and, with some exceptions which we can ignore, every one of those cells contains a complete copy of that body's DNA. **Thus DNA can be regarded as a set of instructions for how to make a body, written in the A, T, C, G alphabet of the nucleotides. It is as though, in every room of a gigantic building, there was a bookcase containing the architect's plans for the entire building.** The 'book-case' in

a cell is called the nucleus. The architect's plans run to 46 volumes in man; the number is different in other species. The 'volumes' are called chromosomes.

DNA is an incredible, almost magical molecule. Because of this, some very astounding powers have been attributed to DNA, like the one above. In this book I will show that DNA does fantastic things, but it does not hold the plans for the human body, or plans for the body of any species; nor is it the control center of the cell. The only function DNA has is to hold and maintain the plans for all of the millions of proteins of all living organisms, and to operate so those plans can be copied during protein manufacture. That's it. That alone is an immense job.

If you asked a vast majority of well-educated people, "What holds the plans for the human body", most people would answer, "Our genes hold the plans for our body and all of its parts". They think our genes form our eyes, brains, skeletons, muscles... Few non-scientists realize that genes are only involved with the construction of proteins. Most biology books and biology professors credit our DNA with the making of all body parts; they must to support evolution. If people found out that genes were not the holder of the human blueprint, nor that of all species, what would that do to the foundation of evolution? It would vanish in a flash. Hence, evo-illusionists must fool the masses into thinking that DNA is the *god molecule*. That it directs the formation our entire bodies, and all of its parts; and the bodies and structures of all living organisms. *Discover Magazine* has this to say about how our bodies form:

Consider, for example, your body's crowning glory, your head. How did the bone cells in your skull know enough to marshal themselves into a dome, while those in your jaw formed a trap-shaped mandible? And how did those on your left side arrange themselves in a mirror image of your right, and how were holes left in just the right places for your eyes? Come to think of it, how did your head wind up at the top of your body, nodding sagely at the end of your neck and spine? ...How did that blob-like egg give rise to, say, nerve cells in your fingertips with long filaments to cable signals back toward the spine? What informed them to relay their signals to intermediary nerve cells thereby zapping them with chemicals? What told those cells' filaments to twine themselves into a cord inside your spine, levitate toward your skull, and send messages into the folds of your brain for interpretation?

Of course the answer given by Discover to the puzzle of what forms our entire body is... DNA; our genes. Discover said the beginning of the understanding of how our bodies form from just a single fertilized cell came with:

...the discovery in 1953 of the structure of DNA, the genetic blueprint at the core of our cells. That led to quick progress in understanding how genes might direct the growth and fate of cells in a developing embryo.[1]

Genes called homeobox genes or HOX genes are credited with forming all of our body parts. It may be that genes, when incorrectly triggered, can cause mutations in the formation of living embryos, and in specific parts of those embryos. Unneeded multiple parts can form, as well as malformed parts. Mal-coded proteins can trigger mutations. But the notion accepted worldwide that our genes form all of our body parts is simply not possible. Genes are nothing but a cog in a wheel of many biochemical parts that cells use to make proteins. That's it. They do not, and cannot form body parts. Therefore, errors or mutations in gene copying over millions of years cannot have ever formed every living species or the body parts of every living organism on Earth. This fact must be hidden by evo-illusionists, because it kills their theory: *The Theory of Evolution.*[2,3]

The history of science is a fascinating endeavor. Just imagine what Edwin Hubbell must have been thinking when he discovered that the universe was hundreds of billions of times larger than every astronomer in the world thought. Unimaginable. Or when Robert Hook looked in his new microscope and, for the first time in human history, saw the individual cells that we all are made of. Can you imagine yourself present at these discoveries, and how unbelievably exciting that would have been? It's kind of fun to imagine what Hubble and Hook must have been thinking in the aftermath of their discoveries. Solving the puzzle of inherited traits by figuring out the structure and *modus operandi* of DNA was a historically exciting scientific moment. Just think how unbelievable it is that the scientific history of DNA started during the time of the American Civil War, when biological lab equipment was so rudimentary, and the knowledge scientists had about the inner workings of cells was so miniscule. Scientists of the 19th century thought cells were pretty simple building blocks for living structures. They were thought of as more like simple cushy microscopic water balloons. They certainly were not considered to be far more complex than any man-made device of the day, which they were. In fact a single skin cell is more complex than any manmade device of today; far more complex than even the space shuttle. In 1839 Matthias Schleiden and Theodor Schwann proposed the *cell theory*, which states that cells are the smallest and most fundamental unit of life. Their pronouncement was followed by Ernst Haeckel's "discovery" that cells were nothing but "homogeneous and structure-less globules of protoplasm." Boy, was he wrong.[4]

For centuries scientists realized traits and characteristics were transmitted from parent to child. When animals, which includes humans, reproduced, their offspring had many of the same characteristics as the parents. The *modus operandi* of this transfer was a complete mystery. Theories abounded about how this transfer of information took place from parent to offspring. One idea was that animals contained miniature replicas of themselves in the reproductive organs of males. Another idea was called *pangenesis*. Pangenesis theorized that each and every body part sent what were called *gemmules*, miniscule representatives of themselves, to the reproductive organs, which then determined the appearance, and shape of the body parts in the offspring of the parents. In the mid nineteenth century scientists began focusing on cells as the potential carrier of inheritance. Maybe cells weren't as structure-less as was previously thought. Maybe there was more to those seemingly structure-less globules of protoplasm.[5]

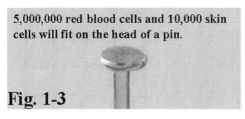

5,000,000 red blood cells and 10,000 skin cells will fit on the head of a pin.

Fig. 1-3

The fact that each individual cell in our bodies is more complex than the space shuttle is inconceivable, since each tiny skin cell or liver cell is so small that 10,000 will fit on the head of a pin. The unwinding of the mystery of cells and DNA is an incredible story. Why did the great people that solved the *Puzzle* of the biochemistry of living cells have the inclination to do so? That is possibly the greatest mystery of all. I am truly inspired by these men and women and what they did. The amazing thing is they were working with internal cellular machinery that they never were able to directly view. Everything they uncovered regarding the molecular structure of DNA and its functioning was due to the results of experiments in the macro-world. "Let's see. If I do this, and the stringy goo in front of me does that, well then it must be composed of ribose sugar molecules..." "If we run an X-ray beam through the goo, and the resulting image on a photographic plate gives a certain pattern, then the molecule must be helical..." Astoundingly, molecules weren't known to exist for certain until the very early 20th century. All of what we know about DNA, and all cellular biochemistry for that matter, is a result of thousands of experiments, and putting results together to figure out the shape and function of unseen and unviewable molecules. The people that did these experiments and thought out the consequences of the results were geniuses at solving seemingly unsolvable puzzles. I feel that if we are discussing DNA and how it works, we cannot eliminate the history of its discovery as a part of that discussion. The next two chapters are a brief history of the uncovering of the mystery of DNA and the demolition of the previous notions about cells; that they were just blobs of protoplasm.

Two unlikely researchers, James Watson and Francis Crick were the first to cross the finish line in solving the structure of the mysterious DNA molecule. What they actually did was put the icing on the cake of DNA discovery. The solution to the DNA puzzle was, in reality, a relay race that took a century to complete. Without the runners that came before Watson and Crick, the finish line would never have been crossed. Watson and Crick didn't actually discover DNA. They didn't make the cake. Actually, and amazingly, DNA was discovered decades before Watson and Crick were even born. But by utilizing the work of scientific pioneers that preceded them, Watson and Crick were able to come to their groundbreaking conclusion about the structure of DNA in 1953. Once the structure was determined, the figuring out of the *modus operandi* of DNA followed quickly behind. What is really astounding is that Watson's and Crick's work was published in 1953. Only eight years later I was studying DNA and how it functions to make proteins when I was taking biochemistry at the University of Southern California. The full story of the structure of DNA and its tasks was up to date in my lectures, and in my textbooks. Which means only a few years after Watson and Crick figured out the structure of DNA, the information was written in biology and biochemistry books throughout the world. If it was in my textbook and lectures, it was certainly in most others. That is unbelievably rapid dissemination of new scientific information. When I was first studying DNA, I had no idea that its molecular structure and *modus operandi* was figured out only a few years before. Thinking it was old information, I wasn't as astounded as I should have been. I studied to memorize data and get good grades so I could be accepted into dental school. At the time, I was unimpressed by the workings of DNA. It represented just a large chapter in an immense biochemistry book. I had to memorize the steps in the process of protein synthesis so I could pass my tests. Now when I study it, I am amazed. Flabbergasted. How could nature come up with such an incredible system? What was once a tedious subject and hard work for me is now a pleasure to study and write about; nothing but fun, and fascination. I hope this book will show you how fascinating the functioning of cells and their most important biochemical, DNA, really is. To me it's a mind-boggling story.

Chapter 2

The Search for DNA Begins

Biology is the study of complicated things that have the appearance of having been designed with a purpose. - Richard Dawkins

Fig. 2-1

The first hints that there was a mysterious but very mathematical underpinning to the inheritance of traits and characteristics from one generation to the next came when in the mid 1850's Gregor Mendel, our leadoff runner in the race to solve the mystery of genetic inheritance, (Fig. 2) began experimenting with garden peas. Pea plants are easy to cross; to mate in a controlled way. This is done by transferring pollen from the *anthers* (male parts) of a pea plant of one variety to the *carpel* (female part) of a mature pea plant of a different variety. To prevent the receiving plant from self-fertilizing, Mendel painstakingly removed all of the immature anthers from the plant's flowers before the cross.

He found that if he crossed green peas with yellow peas, the next generation plants had all yellow peas. What happened to the greens? He next matched yellow peas with yellow peas, and, to his shock 25% of the next generation of peas were green, 75% were yellow. What an unbelievable puzzle! Apparently, the second generation of yellow peas had some mysterious entity in them that held the plans for green peas. Mendel named the yellow trait *dominant*, and the green trait *recessive*. But what was the entity in the plants that gave them the ability to make 25% green peas when both parents were yellow? Because the split was exactly 75/25, Mendel knew that whatever it was in peas that carried traits must have a purely mathematical basis. Mendel's puzzle would take almost 100 years to figure out. But Mendel knew he was on to something really earth shattering. Unfortunately for him, he died before the puzzle was solved. Mendel passed his baton in the race to figure out genetics and inheritance to a Swiss chemist, Johann Friedrich Miescher. [1]

Fig. 2-2

At the time of the Crimean War in the mid-1860's fought by England and France against Russia, Swiss chemist Johann Friedrich Miescher (Fig. 2-2) ran experiments to try to determine the key components of white blood cells, which are part of the body's immune system. The material for this research was easily available as all Miescher had to do was visit nearby hospitals, which were tending to the many wounded Crimean War soldiers. It's unimaginable what it must have been like for those soldiers. Without antibiotics and with only rudimentary anesthesia, treatment must have been a living nightmare. Unfortunately for the soldiers, but fortunately for the march of modern science, Miescher was able to garner a plentiful supply of bandages that were soaked with blood, pus, and notably white blood cells. Fritz, as he was known, carried armfuls of soaked bandages back to his laboratory.[2]

He did many blind experiments on the white blood cells, not having any idea what their results would be. He placed white blood cells retrieved from the bandages in saline solutions. He found that when he added acid to the solution with the white blood cells, they would emit an unusual substance which separated out, and which was not protein. The mysterious material dissolved when an alkali was added which means it must be some kind of acid. Until this point in time, it was thought that proteins somehow carried inherited information in our cells. But this new and mysterious material had properties that were much different than the proteins he had experimented with. Miescher called this mysterious material *nuclein*, because he believed it had come from the cell nucleus. Miescher had no real concept of the immensity of the discovery he had made. His work was the first step in a long journey toward discovering and understanding what many scientists consider is the molecular basis of all of life – DNA. Once he realized that he had discovered a new and unknown material that exists in cells, Miescher tried figuring out ways to extract nuclein in its pure form.

The more Miescher experimented with his new material, the more he became convinced it was an important part of cell biochemistry. He was very close to uncovering the elusive role of DNA, or what he called nuclein, despite the comparatively simple tools and methods that were common fare in the 1860's. The microscopes of the day weren't very powerful. The instruments of the day were pretty basic. Unfortunately Miescher was not a good communicator. He was unable to express and promote his discovery to the scientific community. He didn't publish his results until 1874, a full decade after his amazing find. As a result of his reluctance to publish, many years went by before his research was fully appreciated by the scientific community.[3]

Anatomists, physiologists and biologists of the mid 1800's had a pretty good idea how unbelievably complex the human body really was. And that it would take a massive amount of information to direct and code for its development from a fertilized human egg. For many years, scientists continued to believe that proteins were the molecules that held our genetic material. They were sure nuclein simply wasn't complex or impressive enough to contain all of the information needed to hold the plans for the human body. Surely, one type of molecule could not hold all plans and directives for every part of the complexity of the human body? Could it?

Fig. 2-3

Miescher handed the baton off to Walter Flemming, (Fig. 2-3) an anatomist from Germany, who discovered the same fibrous structure within the nucleus of cells. He named the material he discovered *chromatin*. What he had actually discovered is what we now know as chromosomes. Flemming carried his work much farther than Miescher. He correctly observed how chromosomes separate and replicate during cell division, making two new exact copies. He observed how the split chromosomes form the nucleus of each of two new daughter cells. He termed the process of cell division *mitosis* from the Greek word for thread. DNA looks sort of like a thread when it replicates.[4]

Albrecht Kossel, (Fig. 2-4) a German biochemist, took the baton and made great strides in unraveling the puzzle of the molecular make-up of nuclein. In 1881 he was able to isolate and identify nuclein's basic building blocks: *adenine (A), cytosine (C), guanine(C), thymine (T), and uracil (U)*. He determined that nuclein is a nucleic acid, a weak organic acid that contains nitrogen. He gave nuclein its present chemical name, *deoxyribonucleic acid* (DNA). In 1910 he was given the Nobel Prize in Physiology or Medicine, which is very strange. Why not the Nobel Prize in biochemistry? I don't quite get the "physiology *or* medicine" part of his prize, but that's the way it's listed. He sure deserved the Nobel. It's unimaginable how, in 1880 a scientist could come up with such an astonishing discovery.[5]

Fig. 2-4

The early and successful work done by Miescher, Kossel, and Flemming brought a renewed interest in the work of Gregor Mendel. Without supporting scientific discoveries, Mendel's work kind of became background information. Scientists now took a great deal more interest in his ideas, as they seemed to somehow match the work of Miescher, Kossel, and Flemming. A deluge of research was undertaken to try and prove or disprove Mendel's notions of how physical characteristics are inherited from one generation to the next. How oh how could all of this information and all of these clues be put together so they could make some kind of sense?

American geneticist Walter Sutton (left in Fig. 2-5) and German embryologist Theodor Boveri (right in Fig. 2-5) took the baton and began putting genetic clues together so a single theory would match the information that had so far been discovered. They worked to support the idea that the genetic material passed down from parent to child is within the chromosomes. Their experiments helped explain the inheritance patterns that Gregor Mendel had observed decades before. Interestingly, Sutton and Boveri worked independently during the late 1800s but they came to very similar conclusions. Their efforts melded in a perfect scientific union, along with the findings of a few other scientists, to form *the chromosome theory of inheritance*.

Fig. 2-5

Boveri provided the first true evidence that the chromosomes within egg and sperm cells are linked to inherited characteristics. From his studies of the roundworm embryo he also discovered that the number of chromosomes is lower in egg and sperm cells compared to the number in complete body cells. This was an indication that the chromosomes of sperm and egg cells combine to yield a full complement of body cells.[6]

Sutton studied grasshopper chromosomes. He was the first to recognize a connection between chromosomes and Mendel's laws of genetics. Since characteristics occurred in pairs, and offspring received an equal number of chromosomes from each parent and also come in distinct pairs, was it possible they were receiving differing characteristics from separate maternal and paternal chromosomes? The cat was out of the bag. The beginning of the beginning of truly unlocking the mystery of inheritance of traits and characteristics was at hand. But the road would still be a long one.

Sutton's ideas needed to be tested. In 1909 at Columbia University, Thomas Morgan began experimenting with fruit flies. They had rapid generation times, fourteen days, which made them easy to work with. Results could be gleaned in a very short time. Because of Sutton's experiments, and those of other scientists, it was becoming apparent that chromosomes, if they indeed do hold the information to form our traits and characteristics, operated much like an incredibly organized filing system. The nucleus of a cell was thought to be like a large filing cabinet. In the imaginations of these scientists, the nucleus held information in chromosomes, which acted like folders in the filing cabinet. But what is that information? And how does DNA hold it? This must have been an astounding time for these scientists. They were pretty sure they had located the source of the information that was responsible for assembling the human body. They knew *where* the information source for the human body was. But what information was held? Could it ever actually be deciphered? How much information was there? They were so close; but still yet so far.

Sutton logged and charted naturally occurring mutations in fruit flies. These mutations produced characteristics that were not common such as extra malformed wings and white eyes. He zapped fruit flies with mutation-causing substances called mutagens. He noted that some of the resulting mutant traits occurred in groups, associated with each other. This led him to conclude that they were located in close proximity to each other on the fly's chromosomes. He successfully completed a number of experiments that demonstrated that chromosomes have a definite linear order. It was looking more and more like information stored inside of chromosomes was held in a mathematical code. But how could something so unbelievably tiny hold so much information? Step by step, the relay race continued.[7]

Other scientists using microscopes and some very clever technology stained the nuclear material from cells. A series of bands appeared. The material became very brightly colored. For this reason, they named the material *chromatin*. What they were looking at was our chromosomes. Sutton expanded on Boveri's observations through his work with grasshopper testes. He was able to distinguish individual chromosomes undergoing *meiosis,* a form of cell division, which prepares gametes for fertilization. Can you imagine trying to even locate the testes of grasshoppers, much less checking their chromosomes? I bet those grasshoppers were angry as all get out at this intrusion on their testicles. But through these observations, Boveri correctly identified the *male sex chromosome*. We certainly owe a real vote of thanks to those pissed off grasshoppers that donated their testes for the advancement of modern science. I wonder if the grasshoppers volunteered, or were forced... In the closing statement of a paper Boveri published in 1902 he proposed the *chromosomal theory of inheritance* based around principles determined thus far. His theory stated:

1. Chromosomes contain genetic material.
2. Chromosomes are passed along from parent to offspring.
3. Chromosomes are found in pairs in the nucleus of most cells
4. During meiosis (cell division of sperm and egg) these pairs separate to form daughter cells.
5. During the formation of sperm and eggs cells in men and women, respectively, chromosomes separate.
6. Each parent contributes one set of chromosomes to its offspring.

Other scientists discovered that ribose sugar molecules and phosphate molecules were present in chromatin. By 1910 scientists had uncovered all of the most basic molecules that are the building blocks of DNA; a remarkable feat. But how did these molecules fit together? And how did they interact to produce inherited traits? The answer was still decades off.

As I stated earlier, most early 20th century scientists thought proteins carried inherited traits. They didn't realize the connection between proteins and DNA at the time. Proteins and DNA were considered to be completely separate entities. They knew what molecules DNA was composed of. But they didn't have any idea what their structure was, or how they interacted with other biochemicals. One team racing to the solution to the puzzle of genetics was almost disqualified in 1909 when their lead biochemist P. A. Levene (Fig. 2-6) incorrectly stated that there were equal amounts of the four bases, adenine (A), guanine (G), cytosine (C), and thymine (T), within DNA molecules of each cell. He concluded, again incorrectly, that the four bases linked in repeating

Fig. 2-6

order, such as ...ATCG-ATCG-ATCG... Which meant there was no inheritance coding at all contributed by DNA. According to Levene, DNA was basically a long but useless molecule. Levene's idea was called *The Tetranucleotide Hypothesis*. His hypothesis stood as valid and highly accepted for nearly thirty years.[8]

Fig. 2-7

Then along came a scientist named Oswald Avery. (Fig 2-7)

Avery got the relay team back on track. He proved that DNA carried heritable traits from one bacterial generation to the next. Which meant there must be some type of coding in DNA. But where was it? He found that mice that were injected with *dead lethal bacteria* did not die. (Fig. 2-8) And when they were injected with *live non-lethal bacteria*, they lived as well. But when they were injected with *a mix of the dead lethal and live non-lethal bacteria* the mice died. *Live lethal bacteria* were recovered from the dead mice. How did the living bacteria inherit the lethal trait that killed the mice? After much research, Avery and his group of scientists determined that the *dead lethal bacteria* somehow donated their DNA to the *live non-lethal bacteria*. DNA must carry the lethal trait from the dead lethal bacteria to the living non-lethal bacteria. Actually all it would take to disseminate the lethal trait would be for one single non-lethal bacteria to pick up DNA from dead lethal bacteria. Bacteria procreate by cell division at a fantastic rate. So if only one non-lethal bacterium picked up DNA from a single dead lethal bacterium, in a very short time there would be millions of new lethal bacteria. It's astounding that someone would think of running an experiment like this in the first place. Luckily Avery did, and the solving of the DNA puzzle advanced in leaps and bounds because he did. I wonder if the lab techs had a debate over who would handle the lethal bacteria, and who would handle the non-lethal. Hmmm. I would sign up for the non-lethal part of the experiment. See how, piece by piece, the puzzle of DNA was being solved?[9]

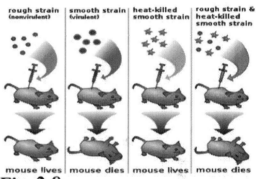
Fig. 2-8

Erwin Chargaff, (Fig. 2-9) a Columbia University scientist, took the baton from Oswald Avery and made great strides toward the

Fig. 2-9

completion of the race. He was fascinated with Avery's work. He began his own experimentation to determine what it is about DNA that allows it to carry the coding for so many traits. He broke down the components of DNA and found that the quantity of adenine (A) always equals the quantity of thymine (T), and the quantity of cytosine (C) always equals the quantity of guanine (G). Chargaff joined Avery in destroying P. A. Levene's theory. Together they probably ruined at least a week's worth of Dr. and Mrs. Levene's dinners. Chargaff also found that the DNA in different species held vastly differing quantities of nucleic acids, but the proportions of A and T, and C and G always remained the same. The fact that the quantities of A=T and G=C became known as *Chargaff's Rules*. Chargaff's findings suggested that somehow adenine (A) and thymine (T) were always connected and paired; and guanine (G) and cytosine (C) were always connected and paired. They also determined that all other possible pairings such as A-C, G-T, A-A, T-T, C-C, or G-G do not occur. Two years after Chargaff's discovery, he explained his findings to Watson and Crick. He handed a very big baton to them. This information was critical when they finally assembled their correct model of the DNA double helix.

Each bit of new information created a new puzzle and a new challenge. Would the correct solution ever be found? At what point would the puzzle actually be considered solved? How deep could they actually venture to do the solving? Would nature put up a wall, and say *this is all you will ever know* as it frequently does? This was of course a concern. Would each new answer, each new advance, each new handing off of the baton, actually be the last nature would allow?

As mentioned earlier, the scientists that were responsible for figuring out what DNA really is, and what its molecular structure is, were two guys named James Watson and Francis Crick. I say "guys" because that's about what they were. They were not the sophisticated scientists that a person might imagine would be the ones to figure out and solve such a complex puzzle. They were unknowns in the field of biochemistry. They had no first-hand experimental data or experience, and they had relatively little knowledge regarding the relevant chemistry. Crick had not yet even finished his PhD degree. Chargaff explained his discovery that quantities of adenine equaled those of thymine, and cytosine equaled those of guanine to Watson and Crick. He had no confidence and no idea that these two would be the solvers of the DNA puzzle. He kind of felt he was dealing with the Abbott and Costello of biochemistry. Chargaff later explained:

I told them all I knew. If they had heard before about the pairing rules, they concealed it. But as they did not seem to know much about anything, I was not unduly surprised. I mentioned our early attempts to explain the complementarity relationships by the assumption that, in the nucleic acid chain, adenylic was always next to thymidylic acid and cytidylic next to guanylic acid...I believe that the double-stranded model of DNA came about as a consequence of our conversation.[10]

Surprisingly, two years later, Watson and Crick, who "didn't know much about anything", solved the puzzle that Chargaff himself was unable to solve. Discovering the molecular structure of DNA became kind of like a space race. The winner would be world famous, and the resultant rewards would be great. By 1950 there were four major teams in the world working on the project. Watson and Crick were not considered as one of these. Several famous scientists including Linus Pauling, who was already a Nobel laureate, headed one. Imposing teams from Cambridge, from Kings College, and another from the University of London were all in the running and working feverishly to unlock the mystery of DNA. Crick and Watson weren't really even considered to be in the race. All four teams had sophisticated equipment that they could use to run tests. Crick and Watson were pretty much in the position of putting together information that was discerned by other people. They actually lost their lab space for a period during the "DNA race", but fortunately got it back. While other teams used their specialized equipment, Watson and Crick used cardboard cutouts and modeling structures that looked a lot like Tinker Toys to make their models and tests.

Chapter 3

Finally a Breakthrough

Trying to read our DNA is like trying to understand software code - with only 90% of the code riddled with errors. It's very difficult in that case to understand and predict what that software code is going to do. - Elon Musk

James Dewey Watson was born in 1928 in Chicago, Illinois. While growing up, Watson was kind of a nerdy kid who enjoyed bird watching. He was a *Quiz Kid* game show prodigy at the age of 15. Obviously he was very intelligent. As a student he studied molecular biology, genetics, and zoology. When he was nineteen he finished his bachelor's degree in biology. Only three years later he was awarded a PhD from the University of Indiana. His PhD studies centered on viral genetics, biochemistry, and radiation genetics. He concluded that new methods were needed to solve the puzzle of DNA, and he wanted to be on the team that did the solving. After he was awarded his PhD, he traveled to Cambridge, England where he had heard some of the most advanced work on DNA was being done. They were using X-rays to study the molecular structure of large biochemical molecules. He wanted to get in on the fun. He was hired by Cavendish Laboratory in Cambridge. Whilst there he linked up with Francis Crick, a theoretical physicist who really knew little about biochemistry. But Crick's mathematical calculations on the structure of large molecules such as DNA were going to make him a key player in the search.[1]

Francis Harry Crick was born in 1916, in Northampton, England. He attended the University College of London and graduated with a Bachelor of Science degree in physics in 1937. Soon he began conducting research toward his PhD, but, in 1939, his path was interrupted by the outbreak of World War II. During that dreadful time, he helped develop radar and magnetic mines. In 1949 he was working at the Cavendish Laboratory in Cambridge where he made the historic link up with James Watson. Watson and Crick began working on the structure of the DNA molecule as a team of two in 1951. Crick did the mathematics and measurement. Watson did the biochemistry. It turned out they were a perfect match.[2]

After only a few months working on the project, Watson and Crick presented their first attempt at making a model of the DNA molecule at a meeting of Cavendish Lab scientists. Their idea was that DNA was a triple-stranded helix with the sugar-phosphate "backbone" as the middle strand. Rosiland Franklin, one of the veteran scientists that attended the meeting, and who had been testing DNA molecules with X-ray beams, contended that, going by her tests, their model was completely incorrect. Further, Watson had miscalculated the number of water molecules held in each DNA molecule. The number of attached water molecules would give a good hint as to what the makeup of the DNA molecule was. Watson's water molecule calculations were way off. Further, Watson and Crick's model had placed the phosphate groups on the inside or middle strand of the three-strand molecule, kind of like inside of a human backbone which is in the center of our bodies. Rosiland

Franklin showed that for the molecule to have as many water molecules as it did, the phosphate groups would have to be on the outside of the molecule, much like on the vertical supports of a ladder. If the phosphate groups were on the outside, they could attract enough water molecules to match Franklin's calculations and observations. Watson and Crick's first demonstration was an overwhelming and embarrassing failure. Further, numerous complaints were posted by other lab scientists that were working on the DNA project. The competition and jealousies between scientists was much like the competition in so many other spirited and competitive events. In a sense it was very much like an Olympic footrace. Only, as soon as the winner finished, the other runners must stop where they are. Every scientist wanted to be the first one to figure out the DNA molecule, and they were worried they would wind up in second place, which might as well be last place. The more qualified and practiced scientists didn't like the competition from these two young upstarts, Watson and Crick. The head of the lab told them to cease all work on DNA. They were taken off of the project. Their lab space was taken from them. Watson was assigned to the study of viruses. Crick was to finish his PhD dissertation.

Fig. 3-1

But these two were not going down without a fight. Their interest in the project was exacerbated with their first failure. They knew the solution to the puzzle was eminent, and they wanted to be the ones who did the solving. As I mentioned earlier, Linus Pauling, already a Nobel Prize winner, and a daunting foe, headed one of the groups that were working on the DNA puzzle. The head of the Cambridge lab was very concerned that Pauling's group was going to beat them to the solution, so the more puzzle solvers, the better. In a short time Watson and Crick were ordered back on the DNA project.

On their second chance at solving the DNA puzzle, they smartly concluded that there were already enough test results to determine the structure of DNA. While other scientists were still trying to glean more information with more and more experiments, Watson and Crick did their tinkering, thinking, and experimenting with the information that had already been provided by earlier scientific studies. Watson and Crick had a copy of Linus Pauling's model given to them by his son. To their relief, Pauling had come up with the exact same incorrect three-strand model that they presented at the meeting, which was soundly and embarrassingly demolished by Rosiland Franklin. So they knew they had more time to work; more time to win the relay.[3-6]

One of the most interesting characters in the cast of characters that uncovered the mysteries of DNA was a Russian scientist named George Gamow (Fig. 3-1). Gamow was primarily a physicist interested in particle physics and astronomy. His radical new theories on the workings of DNA seemed far out of his field of expertise. Gamow was born in Odessa, Russia in 1904. This extremely intelligent scientist worked out a theoretical explanation of alpha decay via quantum tunneling, worked on radioactive decay of the atomic nucleus, star formation, stellar nucleosynthesis and Big Bang nucleosynthesis (which he collectively called nucleocosmogenesis)… and molecular genetics. What a mix. Obviously he was a very intelligent person.

Gamow wanted to leave the Soviet Union due to the severe oppression suffered by the Russian people after the communist revolution in 1918. Gamow and his wife made two attempts to escape Russia in 1932. They tried to paddle a kayak 150 miles across the Black Sea to Turkey. They had to turn back due to bad weather. In their second attempt they tried to kayak along the coastline from Murmansk to Norway, a trip in the Barents Sea of over 120 miles. Poor weather also foiled that attempt. Luckily, in both attempts they didn't die, and the communist authorities did not catch them.

In 1933 Gamow was granted permission to attend the 7th Solvay Conference on physics, in Brussels. He insisted on having his wife accompany him. The Soviet authorities consented and issued passports. The two attended the conference. With the help of Marie Curie and other physicists, they arranged to extend their stay. Over the next year, Gamow obtained temporary work at the Curie Institute, in London, and later at the University of Michigan. They were able to stave off Russian authorities for seven years. Gamow and his wife became naturalized American citizens in 1940 and it was good-bye Russia.

Gamow grew to be very interested in DNA after the discovery of its helical structure. He attempted to solve the problem of how the order of the four different kinds of bases (adenine, cytosine, thymine and guanine) in DNA chains could control the synthesis of proteins from amino acids. Crick has said that Gamow's suggestions helped him in his own thinking about the problem. Gamow suggested that the twenty combinations of four DNA bases taken three at a time corresponded to the twenty amino acids that form proteins. In other words, three of the four code letters of A, G, T, and C would code for each amino acid. The three base pairs are called a *codon*. For example, codon A-C-T might be the code for one particular amino acid. All amino acid molecules are coded in like manner, with three lettered bases. This led Crick and Watson to identify the twenty amino acids that make up proteins. Gamow's contribution to solving the problem of genetic coding was an early guess; a partial win and a partial miss. His idea that three bases coded for each amino acid was ingenious; a huge step ahead in this incredible relay race. It provided an important stepping-stone for the actual discovery of how the genetic code truly functions. But the specific system proposed by Gamow known as *Gamow's diamonds* (Fig. 3-2) was way off the mark.

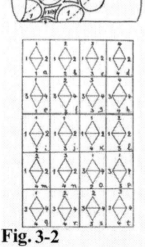

Fig. 3-2

In 1952, Gamow and Watson co-founded the *RNA Tie Club*, a discussion group of leading scientists concerned with the problem of the genetic code. Everyone who attended had to have a tie printed with the latest accepted version of what the DNA molecule looked like. In his own autobiographical writings, Watson later described Gamow's ideas and colorful personality as "zany". He said Gamow was a "card-trick playing, limerick-singing, booze-swilling, practical–joking giant imp". Obviously he wasn't your typical science-guy nerd. Gamow came up with the notion that the nearly identical spacing of amino acids in proteins and of the bases in DNA made possible the direct match-up of amino acid molecules to nucleotide groups on the DNA molecule template. In the drawing he made of his notion of the structure of the DNA molecule and how he thought it functioned, he visualized that the diamonds represent places where the twenty amino acid molecules utilized by living organism would bond to the DNA double helix. The circles represent one of the four bases. Below in the drawing is a code chart that included all twenty amino acids. His thinking was an early step but not close to the actual *modus operandi* of DNA. mRNA, the molecule that transfers protein-making code to a ribosome, a molecular machine shop in each cell, had not even been discovered yet. His thinking was too linear; too focused. His idea only involved DNA. Gamow's amino acid chart was pretty much on the money, though, which made his work enormously important. But he didn't "get the cigar"; or the Nobel. He was not the race winner, but he passed a pretty good baton to Watson and Crick.[7]

Can you imagine what these scientists must have been going through trying to figure out how DNA functioned and what its molecular structure is when they could not at all directly observe a DNA molecule? A single DNA molecule has never been directly viewed by any man, and coming up with its molecular structure and *modus operandi* was a scientific puzzle of grandiose proportions. Gamow gave it a brilliant try, but he was off the mark.

Francis Crick at first liked Gamow's idea, but on further study he dismissed it outright. He wrote a communiqué to the *RNA Tie Club* that stated:

Fig. 3-3

...*there was nothing about either the chemical properties or shapes of the bases to ensure that one and only one amino acid molecule would fit into or attach to the cavities created by a group of bases.*

Watson and Crick traveled to the University of London to work in the King's College lab where scientists were taking the X-ray images of quantities of DNA that I mentioned earlier. They were running tests to see what kind of image results from shooting an X-ray beam through globs of DNA strands. Rosiland Franklin (Fig. 3-3), the scientist that destroyed Watson and Crick's first attempt at a model, was a leading expert at this kind of imaging. To get an X-ray image, wire was stuck in a cork. A thick strand of DNA was hung on the wire. An X-ray sensitive plate was placed behind the DNA. An X-ray tube was aimed at the DNA and activated, which left an image on the plate. Without the DNA target, the image would have just been a solid circle. But the DNA strand diffracted the X-ray beam so that it came out looking like an "X". Fig. 3-4 is the actual image, famously named "Photo 51" taken by Rosiland Franklin that led to solving the DNA puzzle. Franklin's image was far sharper than had been any of the other images taken. Watson wrote that when he first saw Photo 51, "My mouth fell open and my pulse began to race." Photo 51 provided a vital clue that unveiled DNA's double helix structure. The triple-helix structure was officially dead. The number of strands and the precise structure and organization of the individual atoms remained a mystery. There was still a great deal of work to be done and a long way to go. But this was certainly another huge step. Another baton handed to Watson and Crick.

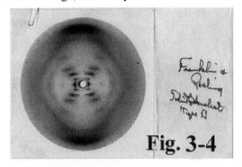

Fig. 3-4

Franklin was correct about her conclusions regarding Watson and Crick's first model. But, despite her own X-ray images, she thought the DNA molecule had a flatter more linear structure like a ladder. Watson got into heated arguments with her when he promoted the helical design. Watson, Crick, and Franklin began this phase of their relationship very much at odds. However Franklin stood high atop the totem pole at the lab because of the way she destroyed Watson and Crick's first model. Where Franklin had been completely right about their first model, her conclusions were very wrong about their second model. Her own X-ray images of DNA should have indicated to her that Watson and Crick were right. She had become a bit too self-assured. She was going to really show them. Instead, her own model that she deduced from her own images turned out to be wrong a second time. Watson and Crick got their revenge with their correct solving of the DNA puzzle. Even so the work she did on attempting to discover the structure of the DNA molecule should have included her in the adulation and rewards received by Watson and Crick. Headlines probably should have read: DNA STRUCTURE DISCOVERED BY WATSON, CRICK WITH HELP OF FRANKLIN. Well, page 737 of Nature Magazine should have made some mention of Franklin anyway. Just imagine how close she was to solving the puzzle, to crossing the finish line herself. She must have gone mad when she realized how close she was, and that she didn't take that one additional step that she could have so easily taken. She had all of the information needed to solve the puzzle right at her fingertips.[8,9]

Franklin's X-ray images and calculations showed that the DNA molecule was *uniformly* about 20 angstroms across. An angstrom is the diameter of a hydrogen atom, or one ten-billionth of a meter. This means the molecule has uniform and parallel outer sides, again, much like a zipper. Imagine a ladder that is uniformly 20 inches across, like most ladders are. Pauling's model, with the backbone in the center, would have had an irregular outer edge. Further experiments determined the number of cross members in the molecule per wind of the helix. These measurements were another clue that DNA is zipper-like.

Watson and Crick were positive they were working with a helical molecule much like a twisted zipper. But they just couldn't get the cardboard parts of the model molecule they had constructed to fit together. Then Crick came up with a major breakthrough. He had been working with protein molecules, and realized many of them were what are called *antiparallel* molecules. That is, the order of the atoms on one side of a ladder-like protein molecule is the exact opposite of the order of the other side. In other words, one side may be in the order:

...-HHPOCC-HHPOCC-HHPOCC-HHPOCC-... whilst the other side would be ...-CCOPHH-CCOPHH-CCOPHH-CCOPHH-...

As you can see, they are composed of identical molecules, only one runs right to left, the other runs to the left to right. (Of course my lettering is only an example.) Picture these being the sides of a ladder. The bases adenine, thymine, cytosine, and guanine made the steps of the ladder. Crick knew that these side strands must be wound around each other. That's the only way they could be aligned, as they were only 20 angstroms apart. If they were structured like a flat ladder, they would have to be wider, or the rungs wouldn't fit between the rails. If you could coil up a 22-inch wide flat ladder, its width would decrease noticeably if you were observing it from a single two-dimensional viewpoint as in a photograph. The 20-angstrom sides of the DNA molecule were simply too close together to follow any other model than a helix. Once these conclusions were verified, the building of the DNA model from the cardboard cutouts made by Watson continued so much more effortlessly; because this was the case, they knew they were on the right track. They had ordered machined Tinker-Toy-like parts to do their modeling, but the parts had not yet been completed by their machine shop. So they continued working with cardboard cut out parts, which fortunately worked well.

Watson and Crick were certain they had a molecule composed of two intertwining helix backbones like the sides of the twisted ladder I mentioned. But how were these held together? The idea of the twisted zipper or ladder had not yet materialized in their minds. Then on February 28, 1953, Watson was sitting at his desk playing with the model bases and double helix rails. He tried fitting the cardboard model A, G, T, and C bases between the double helix sides. Some ways he tried caused the cardboard molecule sides to bulge and narrow, which certainly didn't fit the consistent 20-angstrom model gleaned from Rosiland Franklin's X-ray tests. Then he had one of those moments that was simply exhilarating. He matched up the cardboard models of adenine (A) with thymine (T), and cytosine (C) with guanine (G) and found that the length of A and T equaled the length of G and C. Exactly. And fitting these into the helical sides of the molecule gave him a perfect fit. A fit just like a twisted ladder! Of course each rung was made up of two parts, A and T, and G and C. All of the tests, and all of the data fit this model exactly. It explained why the number of A molecules always equaled T's and the number of G's always equaled C's. What a find! What a feeling that must have been for them.

Interestingly, in 1951, Franklin had come up with an amazing collection of data and mathematical measurements that were sufficient to solve the DNA puzzle. In the autumn of that year she naively gave a lecture at King's College where she revealed all of her data. She just didn't realize what she had given away. Strangely, James Watson was in attendance. It was rumored that he was so distracted by her "looks and sense of dress" that he didn't take the notes he should have. Was he infatuated? Had he listened to her lecture more intently, and taken notes, he could have provided Crick with the vital numerical evidence needed to solve DNA a year before the breakthrough finally came.

Franklin had the data needed to solve the DNA puzzle, but she didn't know what it actually meant. She didn't know what do to with it. Can you imagine her anxiety? So close to the solution, but yet so far. Her data showed that DNA was configured in two chains, each in some way matching the other. The way the component nucleotides or bases on each strand were connected meant that the two strands were complementary. This allowed for the possibility that the molecule might make copies of itself so its information could be passed on to daughter cells. Her laboratory notebooks reveal that she found it difficult to interpret the consequences of the complex mathematics that she had so far accumulated. Crick, with his mathematical

Fig.3-5

skills, quickly realized the significance of her data. It was amazing how Franklin came up with her math using only a slide rule and a pencil. In her notebook Franklin wrote that, "An infinite variety of nucleotide sequences would be possible to explain the biological specificity of DNA". She was able to take a glimpse at the most pivotal secret of DNA: that the sequence of bases contains the genetic code. She was so tantalizingly close to solving the DNA puzzle. She probably would have done so on her own, but she was just a bit too slow and too late. Watson and Crick were first to cross the finish line.

Fig. 3-5 is a very famous photo of Watson and Crick proudly showing off their "Tinker Toy" model of the molecular structure of DNA. They announced their finding in a modest paper titled *Molecular Structure of Nucleic Acids: A Structure for Deoxyribose Nucleic Acid*. The authors were listed as J. D Watson, and F. H. C. Crick. Watson and Crick figured that the outside rails of DNA had a regular sequence, and could carry no coding, while the ladder rungs could have a specific and variable sequence that could carry genetic coding. Crick told people that he and Watson had "found the secret of life".[5,6] Of course what they did discover we now know was the first step to solving the puzzle of the formation of proteins, and the structure of the molecule that holds protein coding information, which carries the basis for traits and characteristics. What they didn't find was the secret for life; nor did they find the secret for what guides embryonic development. They thought they discovered the blueprint for the assembly of all living things. They didn't. Until DNA's code was deciphered, and studies were made on how its code was utilized by cells, no one was sure what DNA really did. Watson and Crick discovered the entity that holds the blueprints for our hundreds of thousands of proteins; and those of every species on Earth. Finding out that DNA only did proteins must have been a big disappointment for the two. In any case, what they did do was truly earth shaking for the fields of genetics, biology, and biochemistry. Even though it only is a tool in the formation of proteins, evo-illusionists have been able to hijack DNA and form the illusion that naturally selected copy errors in DNA have produced every body part, biochemical system, and species on Earth; truly and amazing illusion.[10]

Fig. 3-6

There is one more person in the cast of characters that makes this story so interesting: Maurice Wilkins. (Fig. 3-6) In 1951 he was fascinated with the ongoing race to uncover the mystery of DNA molecular structure. He was one of the first to do studies of nucleic acids and proteins using X-ray imaging. He was able to successfully isolate single fibers of DNA. This is simply unbelievable, since 5 million strands of DNA will fit in the eye of a sewing needle. Can you imagine isolating one of those strands? I certainly can't. Wilkins had already gathered some data about nucleic acid structure when Rosalind Franklin joined the team at King's Lab. Even though Wilkins worked so successfully at the lab before Franklin was even hired, Franklin was placed in charge of all DNA X-ray studies. Wilkins was passed over. He was insulted. Of course he thought he would be, and *should* be in charge of Franklin. Just imagine the strife between the two. Just imagine the dinner conversation he must have had with his wife the day he found out the bad news. "I can't believe it. How on Earth could they... Why those dirty..." His wife might have said, "Have some more potatoes dear..." Wilkins was quiet and reserved. Franklin was confident, outspoken, and certainly not shy. Because Wilkins was told to work for Franklin, and their personalities were such complete opposites, they communicated poorly. If Wilkins and Franklin were able to work together, they might have been the first to discover DNA's structure. Wilkins and Franklin might have been the world famous solvers of the DNA puzzle instead of Watson and Crick. After Watson and Crick published their solution to the DNA puzzle, Wilkins began testing their theory. His testing verified that their double helix and twisted zipper-like structure was correct.[11]

Watson and Crick kept their model at Cambridge. In March 1953, they invited Wilkins and Franklin to come and see it. Upon studying the model, Wilkins and Franklin immediately agreed it must be right; it fit all of the criteria and challenges that they could pose. Everything finally worked. All four agreed it would be published solely as the work of Watson and Crick. They also agreed that the supporting tests would be published by and credited to Wilkins and Franklin. Just imagine the many teams of scientists that were working on this puzzle. The minute Watson and Crick came up with the correct DNA model, the work of the other competing groups had to be trashed. It must have been so very disappointing for all of the scientists who weren't quite able to solve the mystery of DNA structure. All of their lives instantly changed. Watson and Crick became world famous. Linus Pauling already was famous. He probably overdosed on vitamin C. For the losing teams, it was like an instant popping of four balloons. The race was over for them; in a flash. All of those years of effort, and all of the money it took to support that effort evaporated.

Fig. 3-7

In April of 1953, there was a huge party at King's Lab to celebrate the solving of the structure of DNA. Rosiland Franklin did not attend. In 1962 Watson, Crick, and Wilkins were awarded the Nobel Prize for biochemistry. Sadly, Rosiland Franklin died of ovarian cancer in 1958, four years before. She never knew that she had been snubbed by the Nobel selection committee. Many thought she was overlooked because she was a female and that sexism played a role; but that wasn't the case. The Nobel Prize can only be awarded to three people who worked on a particular project. The three most significant representatives are chosen. What a strange rule. Additionally, it cannot be awarded posthumously; also strange. For these two reasons, Franklin was left out of the Nobel Prize award. She never really knew the full extent to which Watson and Crick had relied on her data. If she suspected, she did not express any bitterness or frustration. Until the end of her life, she was very friendly with Crick and his wife, Odile. There is no doubt that she was one of the key runners in the race for the solution to the DNA puzzle. She definitely handed off the baton to Watson and Crick. But she herself was never able to cross the finish line. What a drama![12]

Oh, and one more person on my list of famous and historical people who had a great impact on DNA and its promotion to the status of god molecule is the previously mentioned Richard Dawkins; the person this book is dedicated to. (Fig. 3-7) True scientists did great work in discovering DNA and discerning how it functions. Dawkins promoted the discovery by coming up with some fantastic fables about how it operates and what it can to. He promoted the notion that DNA holds the plans for the human body, that a single DNA molecule could have formed through **Dumb Luck** chemical evolution, and that accidental changes in DNA coding are responsible for the original formation of all of our body parts, and the body parts of all plants and animals. One of the great evo-illusions he promoted is that eyes evolved in 250,000 years, what he says is a "very short time". Yes, he even pretends that he knows this for a fact when his only evidence is some cartoon drawings and made up data by Swedish biologist D. E. Nilsson. There is absolutely no evidence for this 250,000-year theory. It's totally concocted information. This is a perfect example of one evo-illusionist supporting another, which makes their illusions seem more valid. Dawkins tells the fable to schoolchildren who are unfortunately naïve and innocent enough to believe him. Richard Dawkins is truly a great evo-illusionist and promoter of *The DNA Delusion*, and so he must be included in the pantheon of people who influenced worldwide thinking regarding the importance of DNA.

I love this quote from Dawkins:

We are going to die. And that makes us the lucky ones. Most people are never going to die because they are never going to be born. The potential people who could have been here in my place but who will in fact never see the light of day outnumber the sand grains of Arabia. Certainly those unborn ghosts include greater poets than Keats, scientists greater than Newton. We know this because the set of possible people allowed by our DNA so massively exceeds the set of actual people in the teeth of these stupefying odds. It is you and I, in our ordinariness, that are here. We privileged few, who won the lottery of birth against all odds. How dare we whine at our inevitable return to the prior state from which the vast majority have never stirred?[13]

This is a great quote in two ways. One is that I agree with the notion wholeheartedly. I never really thought of death in that way; that we are lucky enough to be here and experience all of the marvels of the Earth and life itself while untold trillions haven't been so lucky. So that is truly a good notion. The other is *the set of possible people allowed by our DNA.* Of course DNA doesn't make people. It makes proteins. So Dawkins gets kudos for the good idea. But mixed in with the good is the bad. He sneaks in his fable that DNA directs the formation of humans at the same time. His second notion is completely wrong, as I will demonstrate.

During the first half of the 20th century science thought life was composed of a mix of matter and energy. Now, with the historic discovery by Watson and Crick, one more very important entity had to be added to matter and energy: *information*. The uncovering of the structure and utility of DNA proved that living organisms had to be replete with information. Life now had to be composed of matter, energy, and *information*. Where did this information arise? Does information arise from nothing? Of course not. Information must arise from intelligence. This is where modern science runs into immense problems. There is no instance where it can be shown that information arises from nothing, or complete randomness. In a universe with no life, as the universe once must have been, a sterile vast array of matter, there was no information, nor intelligence to utilize the information. What was the source of the first information, and the intelligence required to utilize it held by the first entity with intelligence? Information and intelligence are codependent inventions, no matter what their source. One cannot exist without the other. Without intelligence, information is completely useless. Without information, intelligence is useless. When they first came into being, information and intelligence were *new, useful,* and *not obvious*. These are the three criteria for any invention that the US Patent Office demands must be part of any idea for it to be patentable. Information and intelligence would both certainly be patentable under USPTO guidelines. Did human information and intelligence emanate from some other kind of intelligence? What brought about the concept of information and intelligence in the first place? At one time neither existed. What caused the advent of both?

The solving of the puzzle of the DNA molecule was only the beginning of an incredible journey, and the finish of an incredible relay race. Of course, the solving of *how* the DNA molecule was utilized in cells was the next immense puzzle that had to be solved. I have always loved the notion that every time you solve one problem, you create new problems to be solved. The solving of the structure of the DNA molecule by Watson and Crick certainly proves that notion. The solving of DNA only led to many more puzzles that needed to be solved. What molecules interact with the DNA molecule? How is the coding on the DNA molecule read and utilized? What does its code stand for? Does it hold the blueprints for all living things?

What forms our bodies? What directs the formation of an embryo so that it can grow into an entire human from a single cell in only nine months? Did Watson and Crick uncover the *god molecule* that does all of this? In truth, in the next few chapters, I will show that the mystery is exponentially greater than it was in 1954 during all of the celebration of what happened in 1953; during the time all of life's mysteries were supposedly solved. Watson and Crick thought they had found the ultimate secret of life. But the more they and other researchers dug into *what* DNA does, the more they realized they had no idea *how* DNA does what it does; the more they realized they had no idea *how* it originated; or *why*. All they did was discover another incredible apparatus *used* by life. In a way, even though their discovery was historically fantastic, DNA turned out to be a great disappointment for Watson and Crick. It wasn't the secret of life and its blueprint as they had hoped.

There are no words in the English language that can describe the ingeniousness of the designs of nature. DNA and protein synthesis are perfect examples of that ingeniousness. Later in this book I will describe, in fairly simple fashion, how DNA is utilized by every cell in our body to form proteins. The invention of protein synthesis is ingenious beyond comprehension and therefore indescribable in any language. **ID** or **I**ntelligent **D**esign is a completely simplistic attempt to verbally describe what cannot be described. **ID** doesn't include the most astounding part of the origin of nature: *invention*. Invention is far more stunning than is design. It's rarely recognized as such. I fully realize the fact that the mere words in any language cannot be put together to describe the source of nature and it's biological systems. I coined the term **I**ngenious **I**nvention and **D**esign or **IID**, which is my attempt at doing so. It's far closer than the humdrum mind-numbing term **I**ntelligent **D**esign. Do I think, on my say, that the term and notion **ID** should be replaced with **IID**? Of course I do. When something is designed, such as a house, automobile, or computer, the designs arise from entities that were already invented. But can you imagine what an incredible invention DNA really is? The argument about whether DNA was the result of **I**ntelligent **D**esign, or whether it came from **D**umb **L**uck **(DL)** happenstance, as promoted by evo-illusionists, is childish, and absurd. The argument should be whether DNA is an invention, created by an immense intelligence, which all inventions are. Just as our ball and socket joints, digital cameras (eyes), computers (brains), tubes (blood vessels and ducts), servomotors (muscles), and virtually every part of our bodies are inventions. Once they didn't exist, and then they were brought into existence from nothingness. They are all certainly new, useful, and not obvious. And therefore they are inventions. Inventions require intelligence. Which means that the term **ID** doesn't come close to describing the true nature of every invention in nature. **IID** is much closer to reality, and much more difficult for the fables and fable-makers of modern science to deal with.

So now you know the history of the solving the puzzle of DNA, the master code behind the inheritance of traits and characteristics in all living species. Hopefully knowing the background of the solving of the puzzle will give you a greater appreciation of DNA and how it works; and what it can do, and what is cannot do. And if you already had that appreciation, you are officially refreshed.

Chapter 4

OK, Steve, Why Doesn't DNA Hold the Plans For Our Bodies?

It is paradoxical, yet true, to say, that the more we know, the more ignorant we become in the absolute sense, for it is only through enlightenment that we become conscious of our limitations. ------Nikola Tesla

Basically here is how evolution works according to modern science: Think of DNA as a tightly wound incredibly long zipper with digital code printed on each zipper tooth. The genetic code comprises four base molecules designated A, G, T, and C, much like the digital code in our computers uses 1's and 0's, and the Morse code uses dots and dashes. DNA is a strip of these letters, billions of bits of code long. Imagine if a small segment of the code reads:

...*AG*TACGCTT...

When gametes (sperm and egg) are formed prior to fertilization, the code held by their DNA is copied. Visualize that a rare accidental copy error causes the A and G of one of the gametes to reverse order like this:

...*GA*TACGCTT...

Notice the A and G are transposed. This is called a *mutation*. Bad mutations are eliminated through DNA correction processes and natural selection. The good errors or mutations are saved by natural selection. To understand natural selection, visualize that the above error added to or initiated the formation of a tiny bit of a beneficial body part or part of a valuable protein. For example, if the reversal of A and G causes the good accidental formation of 2% of an eye, it will be selected for and saved. Of course it would take millions of copy errors to produce something as complex as 2% of an eye. But this is an imaginary example so let's go with it. Species weakened by bad mutations are killed and consumed far more readily, in this case because they don't have 2% vision. The good errors are copied and added to by additional new good copy errors over thousands of generations. Organisms with 2% vision survive better as they can see at least shadowy images. They can avoid their predators, and see their food targets, if only a little bit. New body parts, and new species build up, generation to generation. The 2% of an eye becomes 4%, and on and on. This process eventually formed all eyes, modern species, organs, and biological systems. In the case of human evolution, a series of species evolved, one after the next, until they became ancient apes and eventually the entire human species. Today our genes, after billions of years of evolution, are made up of 3.2 billion bits of code. As I said, evo-illusionists maintain that this code holds the blueprints for every part of the human body, as well as the plans for our proteins, and they're the control center for all cells. Well, that's what they proclaim and teach anyway.

Fig. 4-1

So is the foundation of evolution valid? Some very recent historic scientific advances and discoveries have laid waste to evolution's foundation. This, of course, brings challenges to the notion that humans evolved from earlier apes in the last few million years. If I'm right, we are back to square one as far as trying to determine the origin of homo sapiens; us. I do realize that the vast majority of scientists in the world are strong believers that humans arose through ape-to-human evolution. It's simply a universally accepted concept. Even *Time Magazine* (figure 4-1) from January of 2006 tells us how man evolved. Its headline says, "How We Became Human". They never ask the biggest question: *Did* man evolve from an ape common ancestor? How could little ole' me prove that the world's scientists and Time are flat out wrong? Actually, it's not that tough. Anyone who can understand grammar school arithmetic can prove it for himself or herself. You will see that you won't need to believe anything I have to say, or what any scientist has to say. You will know for yourself once you see the evidence, drawn purely from the facts that will be presented in the rest of this book. So, open your mind, tighten your seatbelt, and be ready to find out that what you've probably believed about genes being the blueprint for the human body is nothing but an evo-illusion. I will prove beyond a shadow of a doubt that mistakes in gene copying called *mutations* did not form any species or body parts. This is nothing but an illusion that has fooled an amazing number of people, scientists included, for a very long time.

The incredible *Human Genome Project*, the deciphering and mapping of the entire code of human DNA, was completed in April of 2003. American taxpayers alone funded it with over 2.7 billion dollars. Funding and sponsorship from eighteen other countries, and hard work by hundreds of scientists and technicians all over the world were also involved. The Human Genome Project verified that, in our DNA, we humans have about 22,000 genes composed of 3.2 billion *base pairs*, or bits of information packaged in our 23 chromosomes. Strangely, even the finishing of the Genome Project lead to more questions. Scientific American says the genome project shows that we have 3.45 billion base pairs and 35,000 genes. The National Human Genome Research Institute said:

The human genome contains approximately 3 billion of these base pairs, which reside in the 23 pairs of chromosomes within the nucleus of all our cells.

So the total base pair count varies by 450,000,000, which is rather astounding because the Genome Project has been completed. Was it just a massive guess, like guessing how many marbles are in a jar? For the sake of this book, I will accept and utilize the 3.2 billion base pairs and 22,000 genes estimates. Deciphering genomes, even after the Genome Project's multi-billion-dollar study, isn't an exact science. [1]

Interestingly, the genomes of most mammals are of similar size. Mice have 3.45 billion nucleotides, rats have 2.90 billion, and cows have 3.65 billion. One might think that the human genome would be incredibly larger than that of mice and rats, since we are by far the most complex species on Earth and physically so much bigger. But sadly for modern science and evolution, that is not the case. Of course, extremes exist: the *bent-winged bat* has a relatively small 1.69-billion-nucleotide genome. The *red viscacha rat* has a genome that is 8.21 billion base pairs of nucleotides long, two and a half times the size of a human genome. Among vertebrates, the highest variability in genome size exists in fish. The *green puffer fish* genome contains only 0.34 billion nucleotides. The *marbled lungfish* genome is immense, with almost 130 billion base pairs; thirty five times the size of ours. One might think with a genome that size, marbled lungfish would have made computers and gone to the moon long ago. C'mon marbled lungfish, you're not living up to your genome![2]

After the unraveling of the DNA puzzle that I described in Chapter 3, virtually everyone agreed that DNA holds the blueprints for the embryonic formation of the entire human body. We humans had finally figured out the source of the human blueprint and what directs the development of a fertilized egg into a newborn infant. Life was good. Amazingly, the notion that DNA in the form of genes directs the formation of the human body is a better-known yet incorrect notion than is the fact that genes only hold the blueprints for the assembly of *proteins*. Ask any person, including most scientists, "What holds the plans for the human body?" and they will routinely answer "Our genes". Ask them what makes proteins, and virtually no one will have the answer. Ask them to name a single part of the system that makes proteins, and you will get a blank stare. Strangely, what DNA is universally credited with doing, directing the embryonic formation of the human body, it cannot do. Few people know what its true function is. DNA only holds the plans for the formation of proteins. If you asked 100 random educated people what DNA does, my bet is you wouldn't get one correct answer.

The result of the Genome Project was astonishing in two ways. That we humans could actually decipher the human DNA code is certainly one. The other? The Genome Project proves without a shadow of a doubt that there isn't nearly enough DNA code to hold the plans for all of our proteins; which means genes cannot and certainly do not hold the plans for the human body. The human body would take millions of times more coding than do our proteins, even if DNA had the ability to direct its formation. The Human Genome Project was like a nuclear blast that completely destroyed the entire basis for human evolution. Headlines should have appeared everywhere that resembled the New York Times cover article in Fig. 4-2. (My headlines, not those of the NYT.) The Genome Project had earth-shaking news for Richard Dawkins and all evo-illusionists. Just think about the changes that should have been brought to the teaching of biological science all over the world, and the changes needed in millions of textbooks. But there were no headlines. There were no changes in any student textbooks in any location. The most obvious results and conclusions of the Genome Project were and are ignored by evolution scientists, and

The DNA Delusion

Fig. 4-2

by most biological scientists and teachers. The Genome Project is treated as if it's a boon to evolution, instead of the evolution killer that it actually is. It's business as usual![3]

If DNA truly were the holder of the blueprint for the human body, it was calculated that it would need to be the building blocks of at least 2 million genes; probably exponentially more. Just think. Each cell in the human body would need to contain at least two million genes if they truly held the plans for the human body. Cells would have to be thousands of times larger to hold the information needed to form even a basic body part or organ from a fertilized ovum. Try to imagine what humans and all animals and plants would look like if our cells were thousands of times larger; like the size of Ping-Pong balls. If our cells had enough information to form even a very basic part of the human body, we would look something like the lady in Fig. 4-3. Again, the results of the Human Genome Project showed that humans and each of their cells holds only about 22,000 genes, less than one ten thousandth of a percent of the number needed, expected, and hoped for by evo-illusionists. There certainly weren't enough to act as the blueprint for the human body.

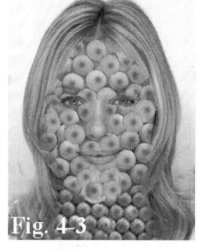

Fig. 4-3

The respected National Institute of Health completely ignored the results of the Genome Project and its effects. As of this writing, the NIH had this to say:

The Human Genome Project (HGP) was one of the great feats of exploration in history - an inward voyage of discovery rather than an outward exploration of the planet or the cosmos; an international research effort to sequence and map all of the genes - together known as the genome - of members of our species, Homo sapiens. Completed in April 2003, **the HGP gave us the ability, for the first time, to read nature's complete genetic blueprint for building a human being.**[4]

Nothing could be further from the truth. It is not possible that the human genome carries the blueprint for building a human being. Is the National Institute of Health being knowingly and completely dishonest here? Genes certainly do play a big part in our biological inheritance. Let me make it perfectly clear that I am not saying they don't. Genes do control most of our *traits* and *characteristics*, such as skin color, height, weight, and hair and eye color, through the proteins, protein enzymes, and hormones they code for. But they do not hold the design plans for the human body, or any body of any organism for that matter. Therein lies the greatest illusion of evolution. Whenever there's an argument regarding whether evolution can form complex body parts, such as eyes, hearts, brains, and skeletal systems, evo-illusionists always morph the argument into whether or not evolution can change traits and characteristics. In the many discussions I have had with evolution biologists, they play dumb, as if they don't understand the difference between complex body parts, and traits and characteristics. The argument usually goes nowhere. DNA and our genes do not control the forming and shaping of the human body, and all of its magnificent parts, from a single fertilized egg. Even so, the notion that *it's all in the genes* is taught in virtually every biology class in schools all over the world. As children that are taught this grow up, the illusion remains with them, so that most adults think the plans for the human body rest in their genes. *It's all in the genes* is an illusion that is so well accepted and pervasive, that, unless this book, or another book on the subject, sells billions of copies, it will probably remain untouched and unchanged for decades, maybe centuries, to come. Writings that uncover *The DNA Delusion* are few and not easily found.

Most of our genes, which are made up of sections of DNA, are wrapped perfectly and incredibly tightly in the nucleus of every cell in our body. The DNA coding in all cells of a single organism is identical. There is no variation. There isn't a special DNA code in liver cells, another DNA code in brain cells, and another in retinal cells. This means that the total number of bits of code in the entire human body cannot exceed 3.2 billion. The 3.2 billion bits of code in human DNA certainly sounds like an immense amount. Most people work with numbers in the tens, hundreds, or thousands in their everyday life. To us mere mortals, it would seem that 3.2 billion should certainly be enough to code for *the entire human body*, and more. Our 3.2 billion opposing zipper-teeth-like base pairs is a colossal number to us. Few people calculate with such enormous numbers, which makes the DNA illusion an easy sell. I'm going to show you that the 3.2 billion base pairs in our DNA isn't even chump change in comparison to the coding needed to form just one part of the human body: the human brain.

In a PBS documentary on DNA titled *DNA: The Secret of Life,* (April, 2003), a scientist states:

DNA produces the brain, the brain understands DNA, so DNA eventually comes to understand itself.

Is this true? Does DNA "produce the brain"? Here is the introduction from a peer-reviewed paper on *Genetic Changes Shaping the Human Brain* from the Howard Hughes Medical Institute. The Human Genome Project concluded in 2003. This paper was written fully twelve years later. It obviously ignores the results of the Human Genome Project.

*The **development and function** of our brain is governed by a genetic blueprint, which reflects dynamic changes over the history of evolution.*

The paper goes on to attempt to explain how changes in the genetic code brought forth the evolutionary development of the human brain. Of course I am going to show that that is not possible, and all of the effort the writers of this paper put in was a complete waste.

The human brain is made up of approximately 100 billion *neurons* or brain cells. Each neuron is connected to the *axon terminals* of tens of thousands of other neurons by branches that emanate from the body of the neuron called *dendrites*. (Fig. 4-4) These connections are much like solder and wire connections on an electronic circuit board. This means that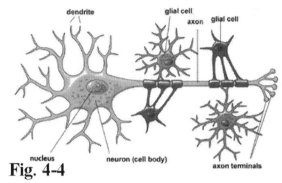

Fig. 4-4

there are *quadrillions* of interconnects between brain cells in the human brain. A single firing brain cell might communicate with thousands of others in a fraction of a second. No computer on Earth is remotely close to the complexity of these communicating bits of living organic matter. In the drawing, for clarity, only a few dendrite branches and axon terminals are shown. In reality, each neuron has tens of thousands of them; some as many as 100,000. There are so many that that it wouldn't be possible to make an accurate drawing of even one brain neuron.

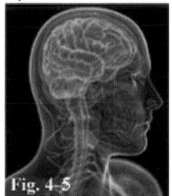

Fig. 4-5

To make things even worse for the illusion that our genes hold the blueprint for the human body, each brain cell requires at least 3 or 4 *glial cells*. (Fig. 4-4) Some sources estimate as high as 10 glial cells. Glial cells provide structural support, protection, resources, and control of *neurotransmitters* (brain biochemicals that send signals from one neuron to the next) for the neurons that make up our brain. Glial cells are kind of the support troops for brain cells. Glial cells and neurons work together to give us all of our brain functions. Glial cells and brain neurons must be constructed and wired together perfectly, just like a

circuit board on a computer, or they wouldn't function at all. Counting only the 100 billion neurons, the over ten thousand dendrite connections of each neuron, and the over 300 billion glial cells in the human brain, there needs to be over 1,000,300,000,000,000 (1 quadrillion, three hundred billion) quantities of information to form and properly connect only the brain's neurons and glial cells. On top of that, a blood supply is needed to feed oxygen to every cell in the brain. This means there are billions of blood vessel cells that must be configured into blood vessels so they can feed oxygenated blood to each neuron and glial cell. During the formation of the embryo, incredibly complex plans must exist that can guide the development of each neuron and its thousands of interconnects with other neurons, and with each glial cell. Additionally, the embryonic *formation* of the blood supply to the brain must be directed. Also, the overall size and shape of the brain must be planned and guided. It becomes quickly apparent that there needs to be, in some location somewhere in the original human zygote (fertilized egg/ovum) and in the human body, plans that can hold quadrillions of bits of information that can and will guide the formation and development of the fully formed infant and its brain from a single fertilized egg cell. There are numerous other entities that support brain function that require plans and direction in their embryonic development. But neurons (brain cells), glial cells, and their blood supply are all I need to trump the illusion that genes hold the blueprints for the human body.[5-7]

As far as blueprints for the human body go, the 3.2 billion base pairs in our genes are quickly overwhelmed by just the human brain. Now add the trillions of trillions of bits of code needed to form the rest of our body, and science has a real dilemma. DNA and our genes cannot possibly be the holder and source of the plans for forming even the human brain. So, what is?

To further complicate matters for evo-illusionists, over 98 percent of the human genome is considered "junk DNA". It's non-coding, and has no known usage. Evolution scientists try their best to find uses for junk DNA. Tthey have come up with some pretty amazing ones. But even with new and complex uses for non-coding DNA, non-coding DNA outnumbers coding by about 50:1. This means there are only about 64 million useful bits of code in our DNA. (2% of 3.2 billion) Each amino acid molecule, the building blocks of proteins, requires three bits of code, like A-G-T, which is called a *codon* so it can be correctly placed in a protein molecule chain. Which means there is enough DNA code for around 21 million amino acid placements (64 million divided by 3). Watson and Crick must have shuddered every time they watched their molecule shrink precipitously. Just think, they lost 98% of their incredible DNA breakthrough. What a disappointment the discovery of junk DNA must have been for them. Codons reduced the remaining amount of information held by DNA by two thirds. The amazing shrinking DNA molecule.

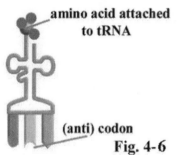

Fig. 4-6

Fig. 4-6 is a diagram of a *tRNA* molecule. Notice the three bases at the bottom that attach themselves to the three-part code or *codon* of mRNA during the manufacture of proteins in all cells. tRNA's are short RNA molecules that carry a single amino acid molecule to the ribosome so that protein molecules can be assembled. If you are not a scientist, all you need to note here is that it takes three coded base molecules at the bottom of the tRNA molecule to hold one amino acid molecule. If we use the minimum estimate for the number of proteins in the human body, 90,000, (estimates run up to 2 million) and multiply by the average number of amino acids in a protein, 500, we need 45 million codons just to form our proteins. Our DNA has only 21 million usable codons. The math quickly wipes out any hope that DNA can hold much more than the coding needed to manufacture our estimated 90,000 to 2 million proteins. Even worse for evo-illusion, there is no known mechanism that DNA has to form any part of the human body. DNA is simply a code holder; a magnificent one at that. If DNA had some *capability* to direct the formation of the human body, which it does not, it certainly doesn't have enough coding by light years to do so. The math is simple; so is the proof that DNA is not the holder of the blueprints for the human body. Since the foundation for evolution, that naturally selected random changes in DNA produced all of living nature isn't valid, the entire theory of evolution isn't either. In this this book I label evolution scientists *evo-illusionists*, and what they conjure are *evo-illusions*. From just the information in this chapter you should be able to see why. The farther you read, the more you will understand why I use these terms; and why this book is secondarily titled *Evo-Illusion of DNA*.[8,9]

Chapter 5

The *How's* Drive the Scientists... and Me Crazy

As we look out into the Universe and identify the many accidents of physics and astronomy that have worked together to our benefit, it almost seems as if the Universe must in some sense have known that we were coming. - Freeman Dyson

Later in this book I will discuss *homeobox genes* or *Hox genes*. Modern scientists credit Hox genes with laying out the basic body forms of many animals, including humans, flies, and worms. But it doesn't matter what our genes are named and credited with, they are maxed out by the amount of information needed to just to hold the blueprints for our proteins. Imagine trying to do a project that would require 150,000 *terabytes* (150,000 trillion bytes) on an imaginary computer that has 8 megabytes (8 million bytes) of computing power. One terabyte is 1,000,000 times larger than a megabyte. This should give you an idea why our DNA isn't close to being capable of doing all functions it is credited with. If the information-holding capacity of our DNA is calculated ignoring the fact that 98% of DNA is un-coded junk, and counting both sides of the DNA "zipper-like" coded molecule, the information carried by our DNA totals about 857 MB. One CD-ROM disc can hold up to 750MB. So our entire DNA is only able to hold a little more information than an average CD-ROM. No matter how you look at it, DNA carries a miniscule amount of information compared to the amount of information needed to form a human body. DNA is an illusory *God Molecule* that is given credit for doing things it cannot do. DNA is *The God Delusion* of modern science; evolution's version of the *Invisible Man in the Sky*.[1]

If there aren't even enough genes to make our 90,000 to 2 million proteins, how can DNA make up that deficit? By an uber-ingenious process of gene splicing. Genes aren't conveniently laid out as single continuous stretches of genetic code. A single gene doesn't code for a single protein molecule, as one might think. Nature just isn't that neat; but it is incredibly efficient. It seems like there is always a caveat. To manufacture proteins, scientists now know cells must utilize mixed and matched gene segments from different locations on the target gene. A little from this part of the gene, a bit from that one, and some from that end, will make up the code and ultimately form the full plan for the assembly of amino acids to form a single protein molecule. Multiple gene segments are used to code for multiple protein segments. This is the only way our 22,000 genes can code for 90,000 to 2 million proteins. The mixing and matching of parts of a gene to make a full code for a protein is called *alternative splicing*. Alternative splicing allows a single gene to code for parts of multiple proteins.

Alternative splicing raises serious difficulties for evolution in two very definitive ways. A single copy error or mutation on a gene can potentially affect the code used to synthesize many proteins that depend on that single gene section for their correct construction. Errors in the formation of proteins can be deadly. They have never been found to be good. The notion that a copy error can be *beneficial* to multiple proteins is preposterous. Even if a copy error were good for an individual protein, the odds that it would also be good for multiple other proteins that utilize the same section of the gene during gene splicing would be infinitesimal. Because of alternative splicing and the damage copy errors would cause to multiple proteins, copy errors alone should eliminate evolution as the source of all of living nature. But the illusion lives on that copy errors cause mutations that are naturally selected resulting in the formation of all species, biological systems, and body parts. This is the foundation of evolution; and it has very quietly crumbled.

Secondly, alternative splicing also makes protein synthesis unimaginably more complex than it was thought to be before its discovery. In humans, the biochemicals that form proteins have to search through multiple coded gene segments multiple times to find exposed codes for the several parts that make up a single protein molecule. In other words, the most incredible scavenger hunt is going on in every cell of our bodies at all times. Protein synthesis can now be compared to trying to find multiple needles in multiple haystacks instead of just one needle in one haystack. Then, in perfectly organized fashion, the found coded "needles" must be spliced together in perfect order so that a perfectly assembled protein molecule is the result. The unused non-coding segments of DNA must be chosen and eliminated by some very smart molecules. The illusion that evolution invented and assembled the first protein synthesis machinery was impossible before gene splicing was discovered. When alternate splicing was added to the complexity of protein synthesis, the notion that evolution had invented and assembled protein synthesis moved to the realm of beyond impossible; actually right where it was before. It made evo-illusion immeasurably more illusory. There simply is no possible connection between any kind of evolution and the origination and formation of protein synthesis. You see, the deeper scientists look, the more complex things get, and the worse things look for evolution; the farther scientists are from figuring out the origin of all or any of Earth's living organisms, body parts, and biological systems. I really wonder when evolution science will be able to take a good honest look at itself, and come up with the only conclusion that can be made: naturally selected mutations are not the source of all living nature, since genes don't hold the plans for any body parts or the whole body of any creature. The true scientific source of living nature must be infinitely more impressive than *random mutations and natural selection*.[2-4]

Anthropology has shown that mankind has a great affinity for telling stories and fables to cover a lack of understanding of any natural phenomena. When dogs see events they don't understand, they are wired to tip their heads sideways. (Fig. 5-1 next page) When humans observe phenomena we don't understand, we may also tip our heads like Richard Dawkins, the world's leading evo-illusionist, is doing here. He's the one in Fig. 5-2. But for sure we humans are wired to form fables and illusions to cover the obvious gaps in our understanding. We do so as if we are programmed, just like dogs and Richard are programmed to tip their heads. The ancient Greeks didn't understand lightning, so they came up with a god, *Zeus*, which explained its existence. Who could blame them? Can you imagine what an ancient

Fig.5-1

person that observed lightning might have thought? Scandinavian mythology possesses a fable about *Thor the Thunderer* who was the foe of all demons. Thor tossed lightning bolts and thunder at his enemies. Thor also gave us *Thurs*-day. Modern scientific discoveries have made these ancient attempts at explaining puzzling natural phenomena seem silly. We like to think modern science has brought us way past the need to make illusions and myths to explain the inexplicable. But are we past that programmed need? Have we stopped making fables and illusions for natural phenomena we can't explain? Are we really different than the ancient Greeks, Romans, and Scandinavians?

Fig. 5-2

Modern science is still replete with powerful myths and illusions, much like those of the ancients, but obviously ours are far more sophisticated in their appearance. The *DNA holds the plans for the human body illusion* is only one of our modern fables. Richard Dawkins wrote a book called *The God Delusion* (Houghton Mifflin Company, Boston, 2006) to show that there is no god; that god is just an illusion. I very similarly wrote this book about DNA. My book shows that DNA cannot possibly be responsible for the miracles it is credited with. DNA is, without a doubt a miraculous molecule. What it does is truly astonishing; beyond imagination. But even in the world of *miraculous* and *beyond imagination* there are levels and hierarchies. Above these is the level *impossible*. DNA cannot possibly do what it is credited with doing. Which means both evolution and DNA are credited with doing what they cannot possibly do; not in a trillion years.

The information in the human hard drive of knowledge has increased incredibly. We now know what causes lightning and fire, so there is no need to make fables to explain these. But the world of science is still awash with phenomena we cannot explain. The more scientists learn and discover, the more inexplicable phenomena there is that needs explaining. For example, we are now able to observe the inner workings of cells, and **what** cells do when they interact in multicellular organism. We know that each cell has billions of biochemical molecules that interact in incredible, organized, and coordinated ways. It's obvious that there needs to be an overriding control center to coordinate all the activity inside of cells. In the human body, cells by the trillions also act in organized and coordinated ways; ways that scientists can observe and document. But scientists cannot explain what entity controls the cells and molecules they're observing. So, in a sense, human knowledge has moved the things we cannot understand to new places. Instead of trying to understand and explain lightning and fire, which we've done, we need to explain **how** all the billions of biochemical molecules inside of cells, and all the trillions of cells that make up the human body, are organized and controlled. The truth is, science has no idea what controls the entities it has only recently discovered. The control center for both the inside and outside of cells is just as mysterious to modern science as lightning and fire were to the ancients. Their gods were Zeus and Thor for lightning and thunder. Our *god* that explains how all cells are controlled both internally and externally as teams of trillions of cells, is DNA. Yes, DNA is our Thor and Zeus. Modern man has the same proclivity to make up illusions to explain phenomena that cannot be understood, as did the ancients. In some ways we are only a modern version of them. The only difference is the modern god, DNA, actually can be tested and studied.

Cells that make up the human body, and all living organisms, pose the same problem for scientists that gravity does. Scientists can observe **what** gravity does, and marvel at **what** effects it has on the universe and us, but they aren't privy to understanding **how** it operates and **how** it originated. The greatest minds the world has ever known have tried to figure **how** gravity works, beginning with Sir Isaac Newton's incredible discovery that gravity is a force unto itself. The mathematical formulas he created describe the result of the actions of gravity on matter, but not **how** it operates. Newton described **what** gravity does. He had no idea **how** it functions. **How** do all objects in the universe constantly attract each other? Theorists have proposed that particles called *gravitons* cause objects to be attracted to one another. Unfortunately gravitons have never been detected or observed. Even if gravitons were found, **how** could they pull two objects together that aren't attached in any way? Finding gravitons would create a new problem. We would move the goalposts from, "**How** does gravity function?" to, "**How** do gravitons function?" Science would have to figure out **how** gravitons work.

Another theory is that *gravitational waves* have something to do with transmitting gravitational forces. They're a clue as to how gravity functions. Gravitational waves are 'ripples' in the fabric of space-time caused by some of the most violent and energetic processes in the Universe. Albert Einstein predicted the existence of gravitational waves in 1916 in his general theory of relativity. Einstein's mathematics showed that massive accelerating objects would disrupt space-time in such a way that 'waves' of distorted space would radiate from the source much like the movement of waves away from a stone thrown into a pond. These ripples would travel at the speed of light through the universe, carrying with them information about their cataclysmic origins, as well as invaluable clues about the nature of gravity itself. The strongest gravitational waves are produced by catastrophic events such as colliding black holes, the collapse of massive stars that cause *supernovae explosions*, the colliding of *neutron stars* or white *dwarf stars,* and the Big Bang. These events cause explosions that are so large they would make a hydrogen bomb explosion appear sub-microscopic. Imagine a hydrogen bomb explosion larger than the entire solar system. What is astounding is that an explosion like this would be completely silent because an observer would be in a complete vacuum. If the observer could be close enough to see it happening, he or she would hear no noise.

Gravitational waves were only a theory until 2015 when exceptionally hypersensitive instruments placed at the South Pole sensed distortions in space-time caused by passing gravitational waves. These waves were generated by two colliding black holes nearly 1.3 billion light years away; which means the collision occurred

1.3 billion years ago. The discovery of gravitational waves will go down in history as one of the greatest human scientific achievements. Lucky for us here on Earth, while the origins of gravitational waves can be extremely violent, by the time the waves reach the Earth they are trillions of times weaker and less disruptive. In fact, by the time gravitational waves reached the sensors, the amount of space-time wobbling they generated was thousands of times smaller than the nucleus of an atom. Just think how sensitive those instruments have to be to detect such small distortions. If a collision of the sort that emits gravitational waves were as close as a nearby star, the Earth and all solar planets would be turned into mush.[5-7]

So how do scientists build instruments with such incredible sensitivity? Modern technology is beyond imagination. But again, as always, there is that constant scientific puzzle: *how* does gravity cause the attraction of two separated quantities of matter. Or *how* does it distort space-time? *How* did gravity originate in the first place? *Why* does it exist at all? Einstein advanced gravitational theory by determining that gravity warps space-time; but that doesn't describe what transmits that warping. What is truly astounding is the fact that Einstein came up with the notion that completely empty space has a fabric that can be altered. His thinking was unimaginably intelligent. He came up with the notion that nothing, empty space, isn't really nothing. "Nothing" has a fabric, and a time. What seemed so completely obvious, that space was truly empty nothingness, wasn't.

Gravity is just one of the four forces of nature. These four forces—*electromagnetism, gravity*, the *strong force*, and the *weak force*—are the dominant entities that make matter what it is. In a very obvious way we use two of these forces everyday. Electromagnetism displays itself in numerous ways. Light from the sun comes to us in the form of *electromagnetic waves*. So light is electromagnetism. Our visual system then transforms these waves into the color images we see. Electricity is electromagnetism. The magnetic attraction of two magnets is caused by electromagnetism. Lightning is electromagnetism. Electromagnetism is so varied it would seem that it is a number of different entities; different forces. But all are one and the same: electromagnetism. Of course gravity keeps our feet pinned to the Earth. It keeps the Earth on an elliptical path circling the sun. It keeps the stars in galaxies rotating around their galactic centers. It crunches black holes into the mystical near infinitely dense balls that they are. While electromagnetism shows itself in incredibly varied ways, gravity shows itself to be the comparatively simple attraction of all matter in the universe.

Most people think atoms are somewhat like solid tiny balls. Few have any notion that the forces of nature are as important in shaping atoms as are the particles that make up the atoms. Nature's forces make atoms what they are as much as the particles they are composed of. The positive charge of a proton and the negative charge of an electron are manifestations of electromagnetism. Atomic nuclei are composed of protons and neutrons bunched together. Protons are all positively (+) charged. Neutrons have no charge so they don't attract or repel each other. Just like two positive ends of two magnets repel, the electromagnetic force causes protons in the nucleus of atoms to repel each other. The *strong force* works against this repelling electromagnetic force. (Fig. 5-3) It's one hundred times stronger than the

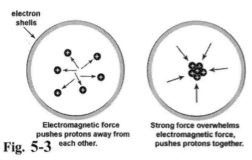

Fig. 5-3 Electromagnetic force pushes protons away from each other. Strong force overwhelms electromagnetic force, pushes protons together.

electromagnetic force, so it's able to force the protons together. If there were no strong force, there would be no atoms more complex than hydrogen atoms in the universe. The entire universe would be composed of hydrogen atoms, free-floating neutrons, and other atomic particles. There would be no stars, no solid matter; and certainly no life. We certainly owe a huge vote of thanks to the strong force.[8,9]

The fourth force has to do with nuclear decay. These four forces are incredible inventions. Why do they even exist? All four couldn't have existed in their present form at the very beginning of the universe, at the Big Bang, or the universe would not have and could not have expanded. Gravity would have kept all of the matter crushed into an immensely massive and energetic but infinitely small speck; a *singularity*. Electromagnetism would have shoved all like charged particles away from each other or crunched oppositely charged particles together. The universe would have been one disorganized mess; a mass of independent particles. Somehow the four fundamental forces had to originate after the Big Bang. Current theories say the four forces were unified as a single force until the temperature of the universe dropped precipitously over 13 billion years ago. They then separated into the four forces in existence today. How did they "know" to do that? Was it just **D**umb **L**uck that they did so? Or was there some sort of cosmic intelligence involved?

The forces were certainly *new*. They were also *very useful* and *not at all obvious*. Can you imagine a universe without gravity? There couldn't be one. But if there could be an Earth and sun, with humans as the inhabitants, and no gravity, we would have to tie ourselves down so we wouldn't just float off into space. The Earth and all planets would be wandering aimlessly in interstellar space. Well, actually, there could be no Earth, no sun, no stars... Can you imagine inventing or even coming up with the notion of these forces if you lived in a universe where no natural forces existed at all? The forces of nature were inventions beyond anything any intelligent person could form in their wildest fantasies. The relationships between these forces are so incredibly unlikely; they do show great design characteristics. If you stretched a tape measure all the way across the universe, and that tape measure represented the strong force that holds protons together in the nucleus of atoms, gravity would represent one inch of that tape measure! Gravity is one trillion trillion trillionth as strong as the strong force. If gravity were two or three inches of that tape measure, we would be slowly crushed against the Earth. A two-hundred-pound person would weigh four hundred to six hundred pounds. Make gravity six inches of that tape measure, that person would weigh twelve hundred pounds. We would be heavy pancakes. Of course, the Earth would have been crunched into a much smaller and far denser ball or, worse yet, gobbled up by the sun's gravity. As I said, the strong force is 100 times stronger than the electromagnetic force, so it's able to keep atoms together. These forces were all established in an incredibly balanced perfect-strength relationship with each other so that nature works and life could form at some future time. Yes, they are inventions that show design. Another question that should be asked: why is there just those four forces? Why not seven forces? Why are there not forces that destroy the "work" of the four forces, that destroy atoms, or that destroy life? The four forces are just too neat, too balanced and perfectly designed. How strange and ingenious that there is a force solely for the purpose of undermining the work of one of the other forces. The strong force is present only to counter the electromagnetic force. It exists only in the nucleus of atoms. Inventiveness, ingeniousness and intelligence must exist in a universe that is supposedly devoid of intelligence. Modern science says the universe exists and is driven by **Dumb Luck**.

In the case of gravity, scientists can observe *what* gravity does to time, matter, and space, but they still have no idea *how* it does so. How did the ingenious invention

Fig. 5-4

of gravity come into existence in the first place? Why does gravity exist at all? The abyss that separates the *what* and *how* of gravity is massive, yet its existence is muddled by the illusions of modern science. Evo-illusionists often cite that science has conquered the *how* of gravity. They claim that it's understood *how* the ocean tides work, just as they know *how* all of living nature came to be. All they need to do is fill in a few details; which is an illusion. The *gravity illusion* is used to promote *evo-illusion*. Evo-illusionists frequently scoff at evolution skeptics by trying to bait and switch gravity with evolution. Gravity and evolution are all just part of modern science, they say. They act as if they have gravity figured out, just as they have the origin of living things and their biological and biochemical systems figured out. In reality, they know nothing about the origin and *modus operandi* of either life or gravity. How can something distort an entity it doesn't touch? How is the force that causes spatial distortion transmitted? Scientists know *what* living nature does, and *what* gravity does, but they have no idea *how* they do *what* they do or *how* they originated. But they act as if they do know. The same is true for the other three forces of nature, electromagnetism, and the strong and weak forces. We are only privy to *what* they do. Not **how** they do it or **how** they originated. Fig. 5-4 is a photo of Richard Dawkins, tipping his head, trying to make up a fable about **how** the four forces of nature function, and **how** they formed; and why. So far, he's had no luck making up a reasonable fable. But I'm sure if we wait long enough, he'll certainly come up with something. I bold the words *how* and *what* because scientists, in the entire history of the study of biological sciences so far, have only increased the knowledge of *what* cells and nature's forces do. In doing so, they made the *how* an even bigger puzzle.

Each time scientists make new and fantastic biological discoveries regarding living cells, nature has presented scientists and its interested observers with the illusion that we have become infinitely closer to understanding the workings of life and those living cells. In reality, the initial excitement of the finds, and nature itself, has presented a massive illusion: the illusion that we are so much closer to understanding *how* cells and their biochemical systems originated and *how* they do what they do. The more we learn about *what* they do, the farther we travel from the understanding of *how* they do it, and *how* cells originated. But the blurred lines between the words *how* and *what* create the illusion that scientists and the people that learn from their achievements are closer to understanding *how* cells function and *how* cells originated. Reality is, the more *what* we know, the less *how* we know. In this venue, *what* is universally inversely proportional to **how**.

Science has promoted a massive illusion that fools people into thinking we understand far more than we do. Exacerbating this illusion is our phenomenal technology. There is no doubt that human technology has advanced beyond anything we thought possible in the 1950's, when Watson and Crick deciphered the structure of the DNA molecule. Can you imagine what Darwin would say if he could observe modern science and technology? He would be stunned. The line between science and technology has become hazy. Most people think the two have advanced equally. In reality, advances in technology have lead to more new advances. Advances in scientific knowledge have given rise to new advances, but have also caused science to move exponentially farther away from solving *The Puzzle* of *how* life and the universe operates, and *how* they originated. All the while, science promotes the illusion it is closer to solving the *how*'s. Of course the biggest question of all is *why*? Why is the universe, Earth, and life existent at all? Science's illusion that it knows the *how's* is why I say modern science still has one foot in the Dark Ages. The modern version of the Dark Ages is still loaded with fables and illusions that try to explain inexplicable scientific phenomena. The illusions are so profound because they are mixed in with modern technology, valid phenomenal scientific discoveries, unimaginable amounts of new information that weren't available to the thinkers of ancient times, and great illusionists who sell the fables such as Richard Dawkins. Unfortunately, modern fables and illusions are far more difficult to detect than were ancient fables.

To give you an idea of how scientific illusions work, when Darwin wrote *Origin of Species by Means of Natural Selection*, cells gave the illusion of being simple little blobs, almost like microscopic water-filled vitamin capsules. The microscopes of the day just weren't very powerful. That illusion followed Darwin to his death. For his entire life Darwin was fooled by the illusion that cells were simple globules, and therefore, they could easily be formed and modified by environmental conditions and changes from generation to generation. Because he was fooled by this illusion caused by the limited scientific knowledge and equipment of the 19th century, Darwin felt pretty comfortable and confident about his theory. Darwin thought evolution created all cells. Scientific knowledge of cells and their incredible complexity and functions increased dramatically in the 20th century. Even so, the illusion that fooled Darwin, that of cells being simple, has continued. The scientific fact is cells are incredibly complex. A single skin cell is more complex than a nuclear submarine. But evolution's illusion is that they are so simple that their parts and functions could have come into being by random changes in their own genetic coding. Yes, you read that right. In this illusion, random changes in the DNA coding *inside* of cells produced... cells!

In the 20th century, the Earth-shaking discovery of DNA and the genetic code gave evolution its *modus operandi*. Evo-illusionists like Richard Dawkins could now make fables about how all living organisms and their organs and biological systems came into existence. Accidental errors in the copying of DNA code during reproduction cycles caused the formation of cells, multicellular creatures, and all biological organs and systems. While Darwin's Theory of Evolution was, in reality, being left in the cold by true scientific advancement, the *illusion* that evolution was the source of all living cells, species, and biological systems remained in place, stronger than ever. Accidental random changes in DNA and the genes it makes up became evolution's illusory source. The overlying illusion used by evolution and by all biological sciences is that everything that makes us human, all of our body parts, and all of our characteristics, are in our genes; in our DNA. It's such a universally accepted fact, it isn't to be questioned. Once Watson and Crick discovered the structure of the DNA molecule and how it functions, every trait, characteristic, and the cellular makeup and shape of every body part of every animal was credited to being held in the genetic coding of DNA. It was said DNA, in the form of our genes, holds the blueprints for our entire body, and the body of every organism on Earth. But does it? Only sixty-four million bits of code in human cells, when quadrillions are needed to hold the blueprints for the human body, says it doesn't.

Fig. 5-5

Each new discovery caused the scientific community and the general population to think science is closer to solving the immense puzzles created by cells and their origin. In reality, each new discovery takes them farther and farther away from solving those puzzles. In light of that fact, I would like to propose a biological law, somewhat like Moore's law regarding storage of digital data. Moore's Law is the observation that:

The future history of computing software will show that the number of transistors in a dense integrated circuit doubles approximately every two years.

The law is named after Gordon E. Moore, (Fig. 5-5) co-founder of Intel Corporation. His prediction presented in 1965 has proved to be accurate. The law is now used in the semiconductor industry to guide long-term planning and to set targets for research and development.[10]

If Gordon E. Moore can make a law, why can't I? I'm going to unabashedly name my law after, ahem, myself. If you think about my law, you will find it fits reality. Somehow my law makes me dream about being famous, at least a bit. If Moore can be made famous because of his law, there is no reason I couldn't be. I can just imagine, eating in a restaurant, and having people come up in droves saying,

"Aren't you Blume? Of Blume's Law?" and "Can I have your autograph?" What a fun thought. Well, I can dream. I even dream this book will sell millions and I will get not only the Nobel, but also the Pulitzer Prize. So here is my law:

Blume's Law Concerning the Origin and Function of Living Cells:
The increase in scientific understanding about what the content of cells and their inner workings are is directly proportional to the distance science is from solving the enigma of how cells originated, and how cells control their internal functions. The increase in scientific understanding about what groups of cells do in multicellular organisms is directly proportional to the distance science is from solving the enigma of how cell groups are controlled and coordinated, how they differentiate embryonically, and how multicellular organisms originated.

Fig. 5-6

The first part of the law is related to individual cells. The second part relates to organs and body parts made up of groups of cells in multicellular organisms. My law states the more science *knows* about cells, their make-up, and inner workings, the farther away it moves from figuring out their origins and what controls their functions. And conversely, the less science *knew* about the inner workings and content of cells, the closer scientists *thought* they were to solving the puzzle of how cells originated and how they do what they do. It's very easy to imagine the simple water-filled capsules that Darwin thought cells were, assembling themselves, and then sticking together to make multicellular organisms.

But as human knowledge increased, and we learned more and more about cells, we also moved farther and farther away from the goal of figuring out their origin and control centers. Science is now light years farther from figuring out the source of living cells; much farther than the scientists of the 19th century *thought* we were. Even so, science presents the illusion that it's closer than ever. If any person on Earth can disprove my law, I will say kudos. But right now it looks pretty good. I'm including my photo (Fig. 5-7) in case things go well for my law.

There's nothing I would love more than if scientists and lab technicians had a way of synthesizing cells, and life. What could be more fascinating? Just think, if you could go to a university biology lab, and watch a biologist make living cells, right before your eyes! Even better, just imagine if you could take a biology course titled, "Bio-Synthesizing Living Cells 101". Premed students might have conversations that would include questions like, "Hey, have you taken Bio-Synth yet?" or "Did you get any living cells in your test tube?" That would really be incredible! In the lab course, week by week, you could follow the formulas given to you by your instructors, or in a lab book, and form different species of your very own laboratory-made living cells. If living cells did form because of countless accidents of biochemicals coming together on the sea floor of a hellishly hot and not-life-friendly Early Earth, as modern science says it did, any biology student or lab tech should be able to make cells under a guided and life friendly environment. Nature has determined that my dream of having science classes and labs that could put together living cells is just that: a dream. But my dream is presented as valid science by evo-illusionists. Cells coming to life billions of years ago from non-living substances is an evo-illusion called *abiogenesis*. It's just another scientific delusion; another chunk of evidence that we still have one foot in the Dark Ages of science.

Molecules jet around the inside of cells at comparatively incredible speeds, as if the molecules themselves are living entities. Each cell in your body makes 2,000 protein molecules per second. To give you an idea of how unbelievable this is, every second our bone marrow cells produce over 100 trillion hemoglobin molecules. Hemoglobin is the protein that carries oxygen to every cell in our bodies. Each cell must also destroy or eliminate 2,000 "old" protein molecules per second or cells would explode in short order because of a bloated oversupply. If you could expand a cell to the size of a watermelon, and see the innards in action, you would view a blur of activity. Science has determined what most of the biochemicals and cell parts that exist inside of cells are, but it has no understanding about what makes them tick. Are the molecules inside of cells some kind of living entities themselves? Do they have eyes? Do they have brains? How do they determine exactly where they need to go, what they need to do, and what their job is, inside of a cell? Do molecules inside of a cell have a self-identity? An awareness? How do they transmit their information all around the inside of a cell? Even more puzzling is, how do the information-carrying molecules inside of every cell transmit their information to the outside of cells of a forming fetus of a multicellular organism, or to the location of any other function or event that requires cells to work in teams of millions or billions of cells? [11,12]

As much as scientists know about cells, they have no idea how their non-living molecules know just where they're supposed to go, just what they are supposed to do, and when they are supposed to do it, to keep the cell alive. Even though brilliant scientists have figured out what most cellular biochemicals are, and their functions, they have no idea how they are controlled and organized. What exactly is the control center of a cell? As far as scientists have come in understanding what cells do, and what biochemical parts are involved, the entity that controls the inner workings of each cell is a complete and utter mystery; a mystery scientists aren't remotely close to figuring out. The fact that there must be a controlling entity in all living organisms is very hidden. Hiding things is what good illusionists do. It should be the greatest and best-known mystery in existence. Everyone should be aware of this great *Puzzle*. We should all be monitoring the success science is having solving it. But there is no success. As my law states, we are going in the wrong direction as far as solving the *Puzzle* of the origin of cells and their control center goes. It would be like the Wright brothers, as their knowledge and aeronautical abilities increased, flying a plane one mile, then one half mile, then not being able to get their plane to go 100 feet, and being completely puzzled as to why.

Just as cells are the building blocks of multi-cellular organisms, *proteins* are the building blocks of cells. Proteins make up *enzymes*, which catalyze (drive) chemical reactions inside of cells. They receive signals from outside the cell wall or membrane and pass it to the inside of the cell. They act as structural units, form scaffolds, and transport other proteins and other molecules around the cell. Hemoglobin that carries oxygen in our blood is a good example of a transport protein. Proteins regulate, form hormones, form tiny electromagnetic motors, help other proteins fold correctly, and make up *ribosomes*, the tiny machine shops that manufacture proteins. Proteins give multicellular organisms structure like the collagen fibers in our skin, bones, and cartilage. There is no doubt that the genes made up from coded DNA control many of the characteristics of humans and all living organisms. Our height is controlled by proteins called *growth hormone 1* and *growth hormone 2*. The color of our skin is determined by *melanin*, another protein. Protein hormones can determine our weight, whether we are heavy, average, or thin. Hair color is determined by proteins. It's a fact that proteins assembled by the coding in our DNA determine many human characteristics, and those of all animals in the animal kingdom. DNA also produces many characteristics in plants; but not all. The color of flowers is determined by their genetic code. The peas in a pea pod can be wrinkled or smooth, yellow or green, also determined by genetic code. These genes can be followed, studied, and mapped which allows scientist to predict what kinds of characteristics can be formed by using different gene types.

What genetic code cannot do is control movement and functions of cell biochemicals. It cannot determine cell type, or control the shape, size, and function of organs composed of cells. It does not control species type. It does not control the development of body parts in the case of animals, or plant parts such as petals, leaves, stems, and trunks. It cannot control the goings on inside or outside of cells, but nevertheless modern scientific illusionists credit it with all of these functions. Proteins must be controlled and coordinated by an as yet unknown control center of some kind. Proteins cannot do the controlling. The coordination and function of billions of protein molecules in each cell is almost infinitely complex. Which means they must be dependent on some other entity for their guidance and control.

DNA and its coding makes up our genes. Genes make up our chromosomes. Our DNA coding is ingeniously organized into bits of information placed in files (genes) and folders (chromosomes). Their organization is almost identical to the way our computer files are organized. The function of DNA is simply to hold and make available the coding used in the construction of our 90,000 to 2 million proteins. But sadly, missing or damaged chromosomes, which are composed of DNA, cause many dreadful illnesses and defects. For example *Angelman syndrome* is a genetic disorder caused by defects in a gene located on *chromosome 15* called the *ubiquitin protein ligase E3A gene*. It causes severe mental incapacity. If these genetic defects occur during fertilization cycles, cells cannot make all of the proteins that the body needs for normal functioning. There is no doubt that defective or missing genes can be the cause of devastating illness. Genes give us our characteristics, and they can certainly give us some terrible problems. But genes do not hold the blueprints for the human body, nor can they control what goes on inside and outside of cells, as scientific illusionists claim.

Under a microscope, animal cells look like tiny little blobs with dark nuclei, just sitting there doing nothing; taking up space. Figure 5-7 is an electron microscope image of a human liver cell. Notice the tightly bound nucleus where most cell DNA resides. To give you an idea regarding the tiny size of most cells, 10,000 liver cells

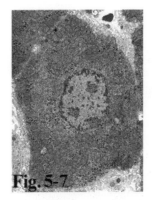

Fig. 5-7

like the one in Fig. 5-7 would fit on the head of a pin. It's very difficult to get a mental picture of what cells do when they're essentially invisible to humans without the aid of a microscope. Cells have an outer cell membrane that holds everything together much like a vitamin capsule contains vitamin gel. Virtually all living organisms are made of cells. Whether an organism is made up of a single cell or is an elephant with trillions of cells, they are still made of cells. The vitamin-capsule-like outer cell membrane holds all of the cell parts and fluids inside the cell. It also keeps deleterious entities out of the cell. Tiny pore-like openings in the cell's outer membrane act as a filter to let specific molecules in and out. Phospholipids (fat molecules) make up most of the membrane structure. Specialized proteins are found around each opening. They act as gatekeepers. They help to select and move molecules in and out of the cell. There are also proteins attached to the inner and outer surfaces of the membrane.

Inside of the cell is a liquid material called cytoplasm, a saline-like solution that holds all the various parts of the cell. Cells have numerous major parts that act as tiny machines that have very specific functions. The functioning parts on the inside of cells operate like the organs that keep humans alive. They are similar to our hearts, lungs, livers, and kidneys. The "organs" inside of cells are called *organelles*. One example is *ribosomes*, or tiny protein-assembling machine shops discussed earlier. Another is the *lysosome*, which breaks down waste materials and cellular debris. Another, *mitochondria*, turns a cell's nutrients into energy. Cells are unimaginably complex building blocks for all living things. Could they randomly originate from random **D**umb **L**uck? If you still think they could, you have to ask yourself "why"? Why would they form from nothing, with no reason for them to do so? An eternity would pass before one living cell would or could form from "nothing". Actually, an eternity would pass before even the notion of a living cell could exist in a sterile universe. In reality the same is true with intelligence. What is the source of human intelligence? In a universe with zero intelligence, what would bring about the invention of human intelligence? Why would it even come into existence in the first place? The universe certainly doesn't need intelligent beings for its existence. Or does it? Does the universe need human consciousness for its existence? Could cells exist without a conscious and intelligent observer present to bring them into existence? Just as life begets life, intelligence must come from intelligence. Otherwise, **why** or **how** would it exist? *How* does intelligence operate? *What* is its source? *What* is the source of the intelligently designed universe? [13,14]

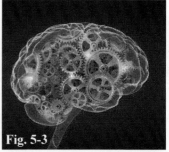

Fig. 5-3

Chapter 6

DNA: The First Great Computer Hard Drive

DNA is like a computer program but far, far more advanced than any software ever created. - Bill Gates

To really understand the earthshaking information in this chapter, and the next, you need to have a basic idea of how our genes, which are composed of DNA, operate inside of our cells. As I said, most people think DNA in the form of our genes is the holder of the plans for the entire human body and every one of its parts. They know police use DNA to solve crimes. But they have no idea what DNA really does. What DNA and its fellow molecular team members do in cells is nothing short of astonishing. I posted several excellent videos at:
www.thednadelusion.com (Alternative: www.thednadelusionblog.wordpress.com) that show what I will be explaining about cells and their DNA. I highly recommend that you take a few minutes to take a look at them. Seeing the processes of DNA is far more astounding when you can watch them in video format. The videos are easy to access on your smart phone, pad, or computer. I will let you know as the subjects in each video come up.

Morse Code		
A .-	B -...	C -.-.
D -..	E .	F ..-.
G --.	H	I ..
J .---	K -.-	L .-..
M --	N -.	O ---
P .--.	Q --.-	R .-.
S ...	T -	U ..-
V ...-	W .--	X -..-
Y -.--	Z --..	
0 -----	1 .----	2 ..---
3 ...--	4 ....-	5 .....
6 -....	7 --...	8 ---..
9 ----.		

Fig. 6-1

The human body uses twenty-two different amino acids, the building block molecules that make up all of our proteins. Our cells can produce twelve of those twenty-two amino acids. The other ten, which are called *essential amino acids*, can only be obtained by eating the right food. So I say eat the right foods![1]

To give you an idea about how codes work, and how DNA coding is far more astounding, technical, and complex than any manmade code, let's take a look at the two most important codes made by humans and feel free to compare. Fig. 6-1 is a chart of the Morse code constructed by Samuel F. B. Morse (Fig. 6-2) and friend. The F. B. stands for Finley Breese, so I see why he went by F. B. Grief actually gave him the opportunity to construct his code. He truly had a tragic early life. In February 1825 his wife Lucretia died in childbirth. Morse was an artist at the time, not an engineer or inventor. Sadly, he was away from home working on a painting commission when he heard his wife was gravely ill. By the time he arrived home, she had already been buried. He wasn't even able to go to her wake. The next year Morse's father died. His mother passed away three years later. He was deeply depressed, as you can well imagine. So in 1829 Morse traveled to Europe, hoping that would get him out of his slump. On his return voyage, in 1832, he met an inventor, Charles Thomas Jackson. The two got into a discussion about how an electronic impulse could be carried along a wire for long distances. Morse immediately became intrigued with the idea. He put together some sketches of a mechanical device that he believed would accomplish the task. Morse developed a prototype of the telegraph. In 1838, Morse formed a partnership with fellow inventor Alfred Vail, who contributed funds and helped develop the system of dots and dashes for sending signals that would eventually become known as the Morse code. So you see, it took two intelligent men to come up with the extraordinary Morse code. For years, the pair struggled to find investors. In 1842 they hit the jackpot. Morse was asked to demonstrate his invention to Maine Congressman Francis Smith. Morse strung wires between two committee rooms in the Capitol and sent messages back and forth. With Smith's support, the demonstration won Morse a $30,000 Congressional appropriation to construct an experimental 38-mile telegraph line between Washington, D.C., and Baltimore, Maryland. On May 24, 1844, Morse tapped out his now-famous first message, "What hath God wrought!" Just think, it took mankind thousands of years to come up with a code that could be transported from point A to point B over a wire. Morse's idea began the information age and changed how the world communicates.[2]

Fig. 6-2

Fig. 6-3

Binary code used by computers all over the world is another code that can be compared with DNA coding. Surprisingly, binary code was an idea that was developed over many hundreds of years. It's not simply a modern code compiled for computers. The ancient Chinese actually invented a rudimentary form of binary coding. In 1623 Francis Bacon created a binary coding system for writing hidden messages. He used the letters *a* and *b* for his coding instead of 1's and 0's. Each letter in in the alphabet had a code. For example the letter L was represented by *ababa*. When text was spelled out in his coding, it looked like a nonsensical jumble of a's and b's. Only the sender and receiver could decipher the code. Of course his coding could be made good use of in wartime situations. He was close, but he wasn't close enough to be christened the father of binary code.

The true father of modern binary code was a German gent named Gottfried Leibniz (Fig. 6-3). I can't believe his hair choice. When I see pictures like this I always wonder if he dressed like that on a daily basis. Anyway, in 1679 he fashioned the first known and charted binary code using 1's and 0's. Leibniz wrote of his system in an article titled *Explication de l'Arithmétique Binaire*. Leibniz was trying to find a way to express verbal statements mathematically. Unfortunately he never really found a useful purpose for his coding. In trying to manufacture a use, he said his method could be used with any objects at all:

The DNA Delusion

...provided those objects be capable of a twofold difference only; as by Bells, by Trumpets, by Lights and Torches, by the report of Muskets, and any instruments of like nature.

Actually, it could be used with anything that is either *on* or *off*. Had he known how important his binary system would become, how it would change the world, he would have been overwhelmed. Can you imagine the conversation he might have had with his wife at the dinner table? "Gladys, I've changed the world. I just started the *information age!*" Gottfried and his wife would have had many happy dinners.

Binary Code

A	100 0001	H	100 1000	O	100 1111	V	101 0110
B	100 0010	I	100 1001	P	101 0000	W	101 0111
C	100 0011	J	100 1010	Q	101 0001	X	101 1000
D	100 0100	K	100 1011	R	101 1010	Y	101 1001
E	100 0101	L	100 1100	S	101 0011	Z	101 1010
F	100 0110	M	100 1101	T	101 0100	a	110 0001
G	100 0111	N	100 1110	U	101 0101	b	110 0010

Fig. 6-4

Binary coding wasn't put into use in rudimentary computers until a graduate student from MIT, Claude Shannon, noticed that binary coding was similar to an electronic circuit. Electronic circuits are either *on* (1) or *off* (0). Shannon wrote his thesis on the subject in 1937. Shannon's thesis became a starting point for the use of the binary code in the computers we have today. Some of the earliest computers were made with transistors, which are nothing but simple electronic on/off switches with no moving parts. They are either on (1) or off (0). Transistors are about the size of a small firecracker. You can see how binary coding can code for all of the letters in the alphabet in Fig. 6-4. Having two transistors *on*, and five *off* represents the letter A. By hooking-up a series of transistors and following binary coding, transistors can store and transfer tremendous amounts of information. As computers became larger and larger, and held more and more information, transistors were simply too large. They greatly limited the amount of information that could be stored. So the silicon chip was born. Silicon chips are simply miniaturized transistors. Our modern cellphones have silicon chips that are equivalent to about 2 billion transistors. Your cellphone works by the

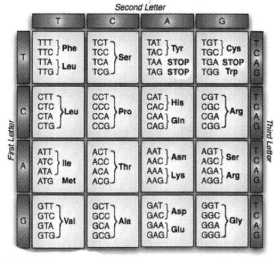

Fig. 6-5

coding of these 2 billion transistors. All of them are either on or off according to their programmed coding. Moore's Law was sure correct about his prediction of computer storage doubling every two years.[3-6]

Both the Morse code and binary code took immense amounts of intelligence to form. Now let's take a comparative look at a far more complex code that scientists say took no intelligence to develop. Of course I'm referring to the DNA coding used to assemble proteins, which are composed of amino acids strung together like the pearls on a plastic pearl necklace. Each amino acid has its own three-digit code called a *codon*. Figure 6-5 on the previous page shows the coding of the twenty most commonly utilized amino acids. They are much like the Morse code and the digital computer code. Notice that DNA coding even has stop codes, TAA, TAG, and TGA, to signify the ends of a line of code. The letters such as Phe are abbreviations for amino acids, in this case *Phenylalanine*.

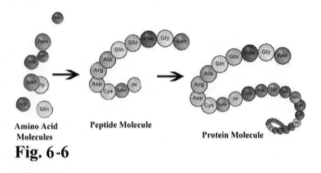

Amino Acid Molecules   Peptide Molecule   Protein Molecule

Fig. 6-6

In protein synthesis, amino acids are connected in chains by a process that starts with the reading of the plans in DNA. As I mentioned previously, if you think of proteins as a snap-together set of plastic pearls, each pearl representing an amino acid molecule, you'll get the idea. Fig. 6-6 shows separated amino acid molecules, which are first joined into a small "necklace" called a peptide. As the shorter *peptide* is added to, it eventually becomes a full long protein chain made up of hundreds of amino acid molecules. Once the full protein molecule is formed, it is bunched up and bent into pre-planned shapes depending on its function. Chains of amino acids that make up proteins are unbranched, just like a pearl necklace. Each amino acid molecule within the chain is attached to two neighboring amino acid molecules, just as in a necklace of plastic pearls. Of course the exceptions are the end two amino acids.

Before you continue reading, I placed a short video, *1. DNA: How It Works*, on my site:

www.thednadelusion.com. (alternate: www.thednadelusionblog.wordpress.com) I highly recommend that you watch it. The video will make everything I write about DNA far easier to understand. When watching it, notice how each molecule does its job like an incredibly intelligent being. Each molecule knows exactly where it must go and what it must do. This video shows a very simplified version of what actually takes place during protein synthesis. In reality, there are thousands of molecules doing very specific jobs needed to make only one protein molecule at any one time. The few molecules shown in the video represent a tiny fraction of those thousands of molecules. The molecules that form proteins do their jobs to near perfection; in fact, far more perfect than could any human perform any task. And they do so without the

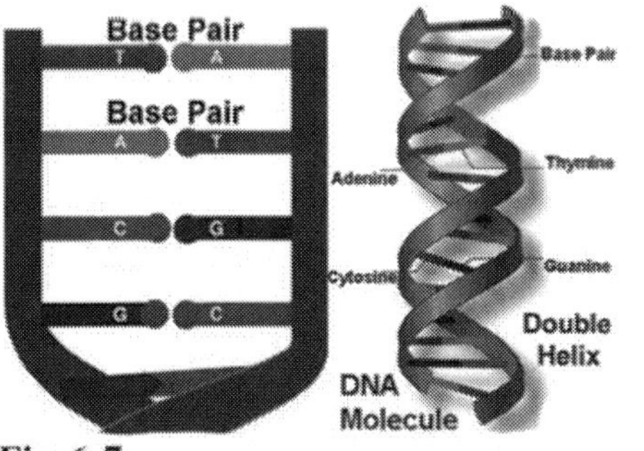

Fig. 6-7

slightest bit of intelligence. They do a flawless job every time. Mistakes come at the rate of one in a billion tries; and most of those mistakes are quickly corrected. Is there anything any human could do that would result in one error per one billion attempts? I sure wish I could play tennis that accurately. But no matter how hard I try...

As you can see in the video, DNA is an immensely long zipper-like molecule that holds a cell's code or template for the construction of all proteins that are used by all living organism. The zipper-like structure is twisted into a *double helix*. The two strands are wrapped around each other as shown in Fig. 6-7. Code is held by the "teeth" of the DNA molecule called *bases*. Note that the teeth in the diagram are paired up much like zipper-teeth. However, they meet end to end, not alternating, to form the code. As I said earlier, there are four kinds of bases or "teeth", A, T, G and C. A always pairs with T, and T with A. C always pairs with G, and G with C. The specific order of these "teeth" forms the code. So, nature was light years ahead of humans in making codes used to construct certain entities and in organizing information by making nature's equivalent of files and folders. Digital computer codes produce images on screens; DNA codes produce proteins. DNA's code is utilized for the continual manufacture of new proteins to replace old ones. DNA molecules are so thin, only 2 billionths of a meter across, that five million strands of DNA could fit inside the eye of a sewing needle. Just think of that: five million! In all human cells, DNA is wrapped very tightly and neatly around small spool-like devices called *histones*. It must be perfectly wrapped up in neat little coiled packages so it can fit inside of the cell. As I said earlier, humans have 3.2 billion paired nucleotide "teeth" in their DNA. If you could unwrap the DNA molecules in a single human cell and put them end-to-end, they would be six feet long. The thought is astounding. Can you imagine trying to wrap a six-foot long thread so perfectly tight, you could fit it inside of a cell so small that 10,000 fit on the head of a pin? To give you an idea of the length and width of DNA, if you could make one cell's DNA molecules two inches wide, and you could place them end-to-end, they would be 29,000 miles long; 4,000 miles more than enough to circle the Earth! So every bit of this incredibly long molecule must be perfectly wrapped to fit inside of a cell that is unimaginably small. Just think. We humans have about 60 trillion cells in our bodies. With each cell holding approximately 6 feet of DNA, each adult human has almost 70 billion miles of DNA! This is enough to go to the sun and return to Earth 375 times.
[7,8]

To give you an idea about the complexity of making only one protein molecule, I've made up a new *musical chairs* game. It's played with fifteen hundred people, and fifteen hundred chairs all lined up and facing the same direction. My game is only for the purpose of showing how complex the making of a single protein molecule can be. Each chair is painted one of four colors; let's say red, orange, blue, and green. Each player is wearing a shirt with one of those four colors. The people with red would have to sit in a blue chair. The people with blue a red one. The people with orange would have to sit in a green chair, and the people with green would sit in an orange one. When the music starts, all fifteen hundred people circle around the chairs. Then the music stops, everyone has to go to his or her correct chair as fast as they can. Can you imagine the confusion by the game players? Oh my gosh what amusement! The uber-confusion is part of the fun of my new musical chairs. Once the music stopped, and all of the people were seated, they would link arms forming a huge snake. They would then sidestep to the outside of the building staying all linked together. They would next go to a small plastic shed with a front and back door like the ones you could buy at Home Depot. Into this scene would walk fifteen hundred new people made up of five hundred groups of three people each. Each three-person group would be holding a giant snap-in plastic pearl about the size of a basketball. There are twenty-two different types of pearls determined by some being larger, some smaller, some ovoid and some round; and many different colors. Each person in each group of three people would have their own colored shirts on; the same four colors as the people in the large group. Each group of three would match their colors to three people linked together in the group of fifteen hundred. If a group of three had, let's say red, green and blue, they would find three people in the long chain with blue, orange, and red as these are the colors that match up to red, green, and blue. They would be standing right in front of them as if they were shaking hands. Each group of three would link up with the other groups of three that have found their counterparts in the long chain. The five hundred groups of three people with the pearls would now be standing face to face with the fifteen hundred people in the long chain. Once the five hundred groups of three people each had correctly located their position facing the long chain of fifteen hundred people, and the two groups of fifteen hundred are facing each other, they would walk sideways through the plastic shed. As they go through the shed, they would snap their plastic pearls together. When they have all come out the other side of the shed they will be Guinness Book of Records record holders with the biggest damn pearl necklace in the world; or maybe the biggest damn musical chairs game in the world. Doesn't this sound like a fun game? I plan on using it at our next picnic. I hope I can get enough people together. Of course my very fun game is completely tongue in cheek. But I made it up to introduce people who are not educated in biochemistry, and those who may need a bit of review, how protein synthesis works. In the next few pages you will see that the chairs represent DNA, and the players represent the other numerous molecules used in making proteins. Can you imagine the mass confusion my game would produce with intelligent humans playing the game? So how do inanimate unintelligent biochemicals play their game 2,000 times per second in every cell in our body? If the nucleotide "zipper-teeth" were as confused as the intelligent humans would certainly be in my musical chairs game, no life could exist on Earth. But nucleotides have no intelligence whatsoever; and no system of propulsion; yet they do their job perfectly. They are never confused in the least. They have no brains to guide them along, yet they do their job far more perfectly than could the fifteen hundred intelligent human musical chairs players and the fifteen hundred pearl necklace makers. The people playing my musical chairs game with intelligence would create complete chaos. The nucleotides with no intelligence whatsoever are smart little molecules, and there is no chaos. Where does that smartness come from? What is astounding is the fact that each cell plays a far more complex version of my musical chairs in every cell in your body, 2,000 times per second. Can you imagine six million human musical chair/necklace makers playing my game 2,000 times per second? Well, that's what each of your trillions of cells do every second of every day.

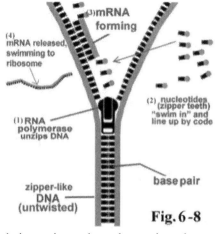

Fig. 6-8

When your cells are forming a protein molecule, it's very similar to them playing my musical chairs game; only in hyper-drive motion. It's like all of the original fifteen hundred people finding their seat instantly as the music ends.

Fig. 6-8 is a diagrammatical representation of the first step of protein synthesis called *transcription*. During transcription the DNA code that is needed to make a protein molecule is copied. The steps are: (1) DNA is unzipped by an enzyme called *RNA polymerase* in just the correct location on the 3.2 billion base pair molecule. How it instantly searches out, on average, a fifteen hundred bit segment on a 3.2 billion bit molecule is beyond imagination. Then zipper-teeth-like nucleotides swim in and attach to the correct coded side of the unzipped DNA. Since there are two sides to this unzipped DNA zipper, how do the nucleotides know which side to attach themselves to? Well, they do, and they never make a mistake. (2) The correct match results in the assembly of a one-sided zipper-like molecule that matches one side of the unzipped "zipper" DNA molecule. (3) The newly formed molecule, called *mRNA*, made up of on average 1500 nucleotides, then is unzipped, and released by another enzyme. It becomes like a half zipper, or one side of a zipper. (4) It next swims to the outside of the nucleus, through a tiny pore in its Saran-Wrap-like outer sheath, as if it's an intelligent snake with eyes. It knows just where to go and just what it needs to do. Once outside of the nucleus, it finds and attaches itself to a molecular device called a *ribosome*, which is like a tiny machine shop, and a bit like the plastic shed in my musical chairs game. (Fig. 6-9) Coded amino acids riding on a tRNA molecule swim in and attach themselves to the mRNA molecule according to their coding. Each amino acid is coded with three bits of code

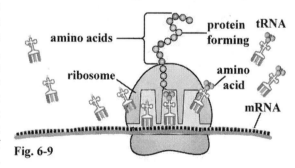

Fig. 6-9

that I discussed earlier, called a codon. These are much like the five hundred groups of three people with the snap on pearls in my musical chairs game. The amino acids are connected like the plastic necklace I mentioned earlier to make the newly assembled protein. Once the protein molecule is fully constructed, it is folded into its utilitarian shape, and off it swims to the location in the cell where it's needed. It too acts just like an intelligent swimmer with swim fins, eyes, and a brain. And, yes, this event occurs in every cell in our body 2,000 times per second. When a protein is assembled in a ribosome from DNA code it is called *translation.*[7-9]

Our cells are incredible ultra-high speed uber-intelligent assembly plants, and few humans, who are composed of trillions of these cells, have any idea or appreciation for them. When reading this information on protein synthesis, did you wonder the same thing I did when I studied DNA? What guides all of the billions of biochemical molecules around the inside of cells so all of these events can take place; so all of the body's 90,000 to 2,000,000 proteins can be assembled? This is a question that is ignored by every textbook and documentary I have ever used to do research on this subject. Every research source I've studied that describes protein synthesis tells *what RNA polymerase,* the enzyme that unzips DNA, does when it comes in and does its unzipping. Then *what* nucleotides do when they attach to the unzipped DNA. Then... They never mention the immense problem of

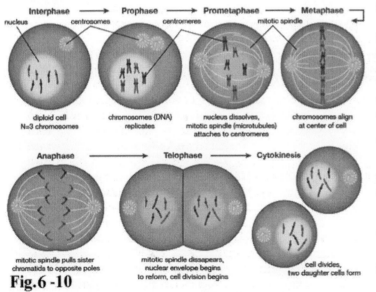

Fig.6 -10

how the biochemical cast of characters inside of the cell moves so ingeniously so they can perform their individual duties. *How* billions of biochemicals are guided around the cell in perfect order is never discussed. It's taken for granted as if it's not an important question. It's like the "given" in a geometry problem. The really crazy molecule is mRNA. *How* it "swims" like a snake that has eyes through a tiny hole in the nuclear membrane, and then "swims" its way to one of millions of ribosome "machine shops", just the correct unused one of course, where it nestles itself in place, certainly is one of the greatest, but most ignored, scientific mysteries known to mankind.

Does mRNA have eyes and a decision-making brain? Do all of the nucleotides and tRNA molecules? What is it that guides all of those molecules into their perfect positions with almost infinite accuracy? It's an ignored question because the illusion persists that biology's god, DNA, guides mRNA and all molecules inside of cells.

Further destroying the illusion that DNA and genes control all cell functions, and hold the plans for all human body parts and cell types, is *cell division*. When cells divide to make two new cells, the parent cell's DNA goes through what is called *replication*. It makes two copies of itself (Fig. 6-10 on the previous page) so each daughter cell will have a full complement of DNA. On my website www.thednadelusion.com (Alternative: www.thednadelusionblog.wordpress.com) is a short video titled *2. Mitosis: Cell Division*. When watching it, notice how each molecular member of the team knows exactly where it must go and what it must do. How do non-living molecules do this? Is there such a thing as "living" molecules and "non-living molecules?

DNA helicase is an enzyme that unwinds *and* unzips all DNA molecule segments during mitotic replication. As the entire DNA molecule is being unzipped, over 6.4 billion nucleotides swim in and align themselves on both sides of the unzipped DNA molecules, forming two new matching "zippers", with 3.2 billion base pairs each. All of this, and the copying of all internal parts of the parent cell occur in from six to eight hours. Just think of what a complex job cell division is for each cell. Their billions of nucleotides have to be perfectly manufactured, inventoried, and controlled so that there are enough nucleotides in the correct locations to make a perfect copy of the parent cell's DNA. The complexity and perfection of what cells do when they divide is unimaginable. They follow the same coding as was used in protein synthesis. In the "unzipped" DNA, A bases attach to T bases, and T bases to A bases. G bases attach to C bases and C bases to G bases. In this way, two new DNA molecules are formed with the same coding as was present in the parent molecule. Two strands of DNA are formed, one for each daughter cell. All the other organelles and entities inside of the cell also make copies of themselves. For example, there are millions of ribosomes in each cell. The cell, in anticipation of full cell division, must copy each one.

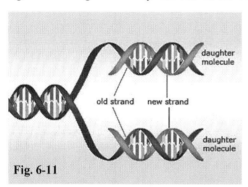
Fig. 6-11

Billions of molecules and organelles must be perfectly copied, split, and then meticulously guided so that each daughter cell has equal parts, and the exact same parts as the parent cell. If the six feet of DNA that is inside of each cell was one hundred miles long, it still wouldn't be enough coding and material to control only one event like cell division. There are innumerable processes that need to be controlled. Even the diagram (Fig. 6-11) showing DNA copying itself is an illusion, since there are billions of parts in the DNA of each daughter cell, and dozens of different enzymes are utilized in the process of DNA replication. DNA polymerase both *builds* and *proofreads* the copy. What an amazing thought. DNA polymerase, a non-living molecule, is capable of proofreading? If so, that means it must scurry around while replication is taking place and do its proofreading. I wonder if I could somehow hire DNA polymerase to proofread this book. Hmm... I bet it would do a perfect job.[10]

Can you imagine if you could watch replication happening in a cell that has been blown up to watermelon size? Actually you couldn't see it happening. There would be a blur activity that would be way too fast for the human eye to see. The typical human chromosome, of which we have 46, has about 150 million base pairs that the cell replicates at the rate of 50 pairs per second. At that rate, it would take the cell over a month to copy one chromosome. The fact that it takes only one hour to replicate a single chromosome is because multiple replication sites are being copied at same time at the rate of 50 per second. Lots of nucleotides from many different points of origin are racing around matching themselves up on the unzipped DNA molecule at the same time. The entire genome of 3.2 billion base pairs is copied in six to eight hours. But remember, since DNA is unzipped during replication, there are 6.4 billion nucleotides to be copied. This means that 222,222 base pairs are copied every second. That's about 222,221 more than we could see if somehow we could watch replication. Can you imagine my musical chairs game being played at this fantastic rate? What kind of cellular control center could control goings on like this? Certainly nothing we humans could ever imagine.

Check out the video *3. DNA Replication: What DNA Does During Cell Division.* on my site:
www.thednadelusion.com (Alternative: www.thednadelusionblog.wordpress.com). All you need to know about the diagram above is that DNA has been unzipped, and nucleotides, the "zipper-teeth" are swimming in on cue to match up their codes. It would be impossible to accurately diagram how complex cell division and DNA replication really are. It's by light-years far too complex. Also, the molecules are far too numerous.

To further complicate matters, the accounting control department of each cell would have to think something like this: "Uh, let's see... We need 6.4 billion nucleotides (bases)...1,986,227,339 A's, umm, 999,228,552 G's, 1,721,277,333 C's, and 1,755,384,001 T's." The 6.4 billion nucleotides alone prove that DNA isn't and cannot be the cell's control center. If DNA were the control center of the cell, it would have to control the molecule type, locations and functions all of the other billions of molecules swimming around the inside of the cell while it's unzipped and completely focused on copying itself. There is only so much a miracle molecule can be expected to do.

In fact, the shape, function, anatomy, position, and number of each cell in our body must be planned and accounted for during embryonic development. When do embryonic cells differentiate into all of the different types of cells needed by the human body? How many times must they divide to form the shape and function of our body parts and organs? When does cell division cease so that the shape of each body part is achieved? How often should each cell divide to produce new cells for each body part? In the active developing embryo, the list of molecules and cells and their shapes, makeup, and functions is endless. Embryonic development must produce trillions of cells, each designed with specific functions and shapes. But 3.2 billion base pairs are all there is in our DNA molecules. You see, 3.2 billion starts looking smaller and smaller... and smaller. As noted earlier, only 64 million of those base pairs are usable. DNA would need quadrillions more base pairs to even have the slimmest of chances of being the cell's control center, even if it truly had a control capability and mechanism. DNA replication itself is more than sufficient evidence against the notion DNA is the control center of the cell. It's just simple math, but with incredibly large numbers, which allows scientific illusionists to continue fooling their audiences.

There are so many other biological functions that cells go through that add insult to injury to the illusion that DNA is the control center of the cell and holder of the blueprint for the human body. Cells have processes that eliminate waste, an unbelievably complex and ingenious system for producing energy called the *Krebs cycle* that turns sugar (glucose) into *pyruvate* so it can be utilized as an energy source, provide storage facilities for molecules, construct miniature roadways called *microtubules* to facilitate the movement of molecules inside of cells, to name only a few. Each additional process adds fuel to the fire that should be destroying the illusion that DNA is the control center of the cell. Unfortunately, the illusion is currently so powerful that it will fool its audiences for decades and probably centuries to come.[11]

DNA in many ways is our modern version of Zeus and Thor. What mysterious entity coordinates and guides each of the billions of molecules so they can move from point A to point B in perfect order inside of cells? What entity directs their manufacture in the first place? What does the accounting so the correct molecules are synthesized and utilized in every case? It's as if every one of the billions of molecules inside of a cell has eyes, a brain, swim fins, and a plan. It's so difficult to give this question the significance and weight that it deserves and that I would like to give it. Words alone won't do the job. Maybe this will help:

How do all of the billions of biochemical molecules inside of a cell move from one location to the next? How do they go exactly where they're needed, and how is this movement controlled? What do they use for propulsion? How does a cell keep track of and direct all of its billions of molecules? Does each molecule have a brain, eyes, swim fins, and a plan? How do these molecules "know" what to do when they arrive at their destination? Is there some sort of unimaginable regulator system that has a sort of remote controlled mechanism that controls all of those billion of molecules?

There. That feels a little better; but not much. Almost all biologists would say that the organization and control of the cell is our DNA, the cell's "control center". In fact, if you Google "What is the control center of the cell?" most scientific websites will say the nucleus is the cell's control center. But as you just found out, all DNA does is hold the code for the making of proteins. It has no special control capabilities over the individual molecules inside of the cell. No known entity inside of a cell does. So many unbelievably complex functions were credited to DNA since its discovery, but as science learns more and more, DNA becomes more and more diminished in importance. It is hugely important as a code carrier for proteins, but that's about it. Each biochemical molecule in each cell swims around like it's an individual organism, completely on its own, but doing exactly what it's expected and supposed to do. When a multicellular organism dies, all the billions of biochemical molecules inside of each of the trillions of cells lose their capabilities that make them seem like they have eyes, brains, swim fins, and a plan. They cease all organized movement and functions almost instantly.

Chapter 7

A Sheep and a Cell Bid *Adieu* to the God Molecule

People think genes are an absolute cause of traits. But the notion that the genome is the blueprint for humanity is a very bad metaphor. If you think we're hard-wired and deterministic, there should indeed be a lot more genes. - Craig Venter

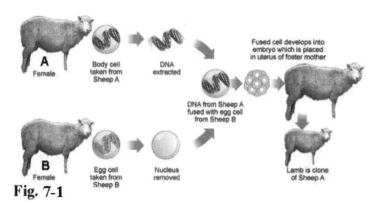

Fig. 7-1

One of the most astounding proofs that DNA doesn't determine cell type occurred in 1996 when Ian Wilmut, Keith Campbell and their colleagues at the Roslin Institute in Scotland performed the first cloning of a mammal. Dolly was a female domestic sheep, and the first mammal to be cloned from an adult somatic (body) cell. Campbell and Wilmut used a process known as *nuclear transfer*, which is the removal of the genetic material of an ovum (female egg) that had half of the genes needed for a whole sheep (B in Fig. 7-1) They replaced it with the genes from a sheep mammary cell that had *all* of a sheep's genetic material (A in Fig. 7-1) Once the ovum had its full genetic code from the mammary cell, it is now called a *zygote*. The new genetically loaded zygote was stimulated to grow by a very small electric shock. It went through several cellular divisions, and became a *blastocyst*, an early embryonic stage made up of 40 or 50 identical stem cells. The blastocyst was implanted in the womb of a surrogate mother sheep. *Differentiation* began, which is the formation of all of the 400 or so different cells needed by the sheep. All the different cell types formed: liver, heart, skin, brain... Dolly developed normally into a fully formed infant sheep. Dolly wasn't just a big boob!

The production of a healthy cloned sheep from only *mammary cell DNA* proved that DNA taken from **any tissue cell in any part of the body** could recreate a whole individual. One would think that a liver cell could only produce liver cells, as they should be programmed to do so. Or in this case, a mammary cell would produce only mammary cells. Somewhere in somatic cells is the information and guidance mechanism to form an entire infant. But where is it? In fact what controlling entity determines that somatic liver cells divide to produce only liver cells, and mammary cells divide to form only mammary cells. Dolly was born on July 5, 1996. She lived until the ripe old age of six. Regarding Dolly's name, Ian Wilmut stated, "Dolly is derived from a mammary gland cell and we couldn't think of a more impressive pair of glands than Dolly Parton's". So now you know the origin of Dolly's name. Wilmut admits Dolly's birth was a lucky accident. He and his colleagues were trying to make clones of fetal cells. They used adult cells as experimental controls—not expecting that they would actually generate a normal embryo. "We didn't set out to clone adult cells. We set out to work with—ideally—embryonic stem cells or things like that," Wilmut said. "Being successful with adult cells was a very considerable, unexpected bonus."[1-3]

If cell type and body design aren't determined by genes made up of DNA, what does determine cell type and body type? Science isn't finished with this enigma, not by a long shot. If DNA was the director of cell and body type, did the internally functioning DNA from the mammary cell "know" what was happening external to itself? Could the nucleus that was transported from the sheep mammary gland to Dolly's ovum "see" its new digs, and make the changes necessary to form a full sheep? The real puzzle is why didn't Dolly's mammary cell start the formation of a sheep fetus while it was part of the mammary gland since its entire DNA was present? Any number of infant sheep should form from just the mammary glands, at any time. What is in the zygote that trumps the forming of only mammary cells in favor of producing an entire sheep? This is just another unsolved and unbelievable puzzle within *The Puzzle*, courtesy of living nature. Scientists thought they had the inner workings of our genetic code all figured out, then along came a sheep named Dolly and the revelation that DNA did not carry all the information that determined cell type and function, and body design. Dolly proved beyond a shadow of a doubt that there must be some other as yet unknown entity that has that job.

Dolly and the Genome Project eliminated the notion that the DNA inside of cells was the control center and plan holder for the designs and functions of every cell, cellular molecule, organ, body part, and characteristic of the human body; or the body of any animal. Humans have only about a third more genes than a nematode worm and only twice as many as a fruit fly while we are millions of times more complex than either. The math is simple. *The Puzzle* of what holds the blueprints for the human body is light-years from being solved. It's actually an exponentially growing enigma; growing as scientists learn more and more about the inner workings of cells.[4,5]

Another earthquake in the rumblings of evolution's illusion is the fact that cells can live without their nuclei. In all complex multicellular animals including humans, the control center for all biological functions is the brain. If the brain were removed from any animal, the animal would die instantly. Scientists thought the same should be true with cells: if the nucleus were removed, since we knew "for certain" that the nucleus is the control center of the cell, the cell should also die instantly.

The cloning of Dolly gave an inkling that cells could live without their nucleus. When the nucleus was removed from Dolly's ovum, it didn't die instantly, as was expected. It lived long enough to allow for the transfer of the nucleus from a mammary cell. Scientists have since removed the nucleus of other somatic cells. They were shocked to find that the cells continued to live and function for several months, just as if the nucleus was in place. Not only did the cells keep on living, they continued to take in nutrition, eliminate waste, rebuild themselves, react to toxins… and do whatever healthy cells do, in completely normal fashion. The cells didn't even seem to "notice" that their nucleus was no longer present. Taking the nucleus out of cells does not alter their behavior in any known way at all, with the exception that they are not able to go through cell division, or manufacture most of their needed new protein molecules. Some DNA resides in another organelle, the *mitochondrion*, which is capable of forming some protein molecules.

Actually, there are two types of cells in the human body that have no nucleus at all. One is called a *reticulocyte*. These are immature red blood cells. They develop into mature red blood cells in human bone marrow. The other cell with no nucleus is a *red blood* cell. Red blood cells ferry hemoglobin, a protein that provides oxygen, to all of the cells of the human body. So, how do RBC's make hemoglobin if they have no cell nucleus? The answer again lies with *mitochondria*. Mitochondria are organelles which are responsible for producing energy in all animal cells. In the early stages of the formation RBC's, mitochondria have a dual purpose. They provide energy and store just enough DNA to code for the formation of hemoglobin. Mature RBC's are composed of 95% hemoglobin by dry weight. Early on the mitochondria are ejected from RBC's so they have no nucleus and no mitochondria. RBC's live for about 120 days before their cell membrane begins to deteriorate. They die off and are eventually eliminated by our digestive system. In fact a large percentage of human waste is comprised of deteriorated red blood cells. Each RBC has 280 million hemoglobin molecules; an astonishing fact. An RBC's hemoglobin molecules in its inner domain move around in incredibly rapid yet controlled fashion, doing just what they are supposed to do at just the right moment, so all human tissues can have the oxygen they need to survive; and they are not directed by a nucleus. What is the control center for those billions of molecules if they have no nucleus to guide them? RBC's must have a control center. The big question is, where is it, and what is it? What an incredible mystery.[6]

Over one hundred years ago, scientist began experimentally removing the nucleus of the largest of human cells, a fertilized ovum. To their astonishment, the enucleated cells went on living, long enough to produce over 40 more cells, and to develop into a *blastula*. Somehow they were able to go through limited cell division without their nucleus.

With modern techniques, unicellular organisms have been enucleated to see what their response would be. I love this dramatic description of the surgical removal of the nucleus of a cell by biologist Bruce H. Lipton PhD:

And now for the big experiment... (Maestro, drum roll if you please.) The scientist drags our unwilling cell into the microscopic operating arena, and straps it down. Using a micromanipulator, the scientist guides a needle-like micropipette into position above the cell. With a deft thrust of the manipulator, our investigator plunges the pipette deep into the cell's cytoplasmic interior. By applying a little suction, the nucleus is drawn up into the pipette and the pipette is withdrawn from the cell. Below the nucleus-engorged pipette lies our sacrificial cell; it's brain torn out. But wait! It's still moving! My God, the cell is still alive! The wound has closed and like a recovering surgical patient, the cell begins to slowly stagger about. Soon the cell is back on its feet (OK, its pseudopods, the "feet" of a cell), fleeing the microscope's field with the hope that it will never see the doctor again.

Following enucleation, many cells can survive for up to two or more months without genes. The viable enucleated cells do not lie about like brain-dead lumps of cytoplasm on life support systems. These cells actively ingest and metabolize food, maintain coordinated operation of the physiologic systems (respiration, digestion, excretion, motility, etc.), retain an ability to communicate with other cells, and are able to engage in appropriate responses to growth and protection-requiring environmental stimuli.[7]

The control center of the cell continues to operate and function even though the nucleus was removed, or when no nucleus exists, as in the case of red blood cells. This further supports the fact that the nucleus cannot be the control center of the cell. So what is? Modern biological science, as advanced as it has become, has no idea. Since science has no notion what the control center of the cell is, it also has no notion what the origin of cells is. *Blume's Law Concerning the Origin of Living cells* is validated. The more we learn, the farther we migrate from even getting an inkling of how cells and life originated.

The real big question, the unanswerable question for all biological sciences is, what force moves and guides all the billions of molecular parts of a cell? Do they have little jet engines that propel them from one location to the next?

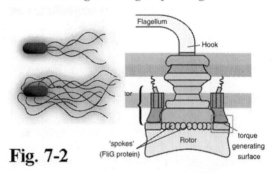

Fig. 7-2

Do all of the billions of molecules think on their own? How do they "know" exactly where they are supposed to go and what they are supposed to do? Do they have a motor and propeller kind of like a bacterial flagellum? (Fig. 7-2 left drawing) You see, some bacteria have one or several corkscrew-like tails that operate like a propeller. An incredibly tiny electromagnetic motor drives them. (Fig. 7-2 right drawing) The motor can operate at 6,000 to 17,000 rpm. These bacteria can move from place to place as they desire. The motor looks like it was engineered by General Motors or Tesla. It can stop and instantly reverse itself at the whim of the bacteria. Do bacteria have whims? Do molecules? If you would like to see the flagellum and its motor in action, go to my website: www.thednadelusion.com (Alternative: www.thednadelusionblog.wordpress.com). Check out *4. The Bacterial Flagellum: An Incredible Electromagnetic Motor.* You will be amazed. Could an engineered motor like this be invented and assembled by **Dumb Luck**? Certainly no human in a million years could put together a microscopic motor such as this. But **Dumb Luck** could? And did?

What entity is inside of bacteria that controls the flagellum motor? Do bacteria have a brain that we have not yet discovered? Do bacteria have yearnings to move from one place to another like any multicellular animal? Do the DNA and the molecules in charge of protein synthesis inside of a cell have a kind of invisible mechanism, like that of the flagellated bacteria, that allows them to move from point A to point B? Do they have little computer-like brains that tell them where to go and what tasks to perform?

Scientists have found an astounding group of proteins called *motor proteins* (Fig. 7-3), that move molecules and entities around the inside of cells. Think of motor proteins being much like "micro-people" carrying immense loads, loads that are twenty times bigger than they are, whilst balancing on a high wire like a wire walker. They have two "legs" and "feet" like we do, and they walk like we do, putting one foot in front of the other. Motor proteins take about one hundred steps per second, obviously much faster than do humans. They walk along *microtubule filaments,* which are like molecular roadways laced in every direction inside of cells. If you visualize a molecular sized version of our own freeway systems, you will get the idea. Motor proteins transport molecular cargo, mostly from the center of the cell towards

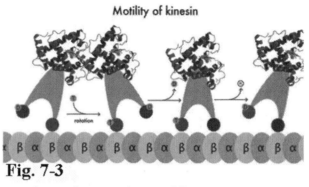

Fig. 7-3

the periphery. The DNA of mammals encodes more than 40 different types of motor proteins. You can see a video of a motor protein in action at my website, www.thednadelusion.com (Alternative: www.thednadelusionblog.wordpress.com). Go to video *5. Motor Proteins*: *Incredibly Tiny But Efficient Movers of Protein Molecules*. Remember, when this little guy is walking his immense load, he is moving at about one hundred steps per second. So this video is in uber-slow motion. The video is only forty seconds long, and well worth the look.

Of course I always wonder how evo-illusionists can fool audiences into believing evolution could have formed these amazing mechanisms. Did one leg and foot evolve first, then the other? If so, did motor proteins hippity-hop on one leg with their giant loads until the other leg and foot evolved? Did both legs evolve at the same time? Is that possible without guidance? How did evolution through natural selection make such an ingenious device? What controls these tiny machines? Do they have brains of their own? Can they see where they're going? How do they "know" which molecules they need to move, and when and where they need to move them? They don't have muscles, so what is their means of movement? Biochemists say motor proteins utilize the energy of *adenosine triphosphate* (ATP) *hydrolysis* to move along microtubules. ATP is the energy source for cells, kind of like gasoline is the energy source for cars. The explanation is kind of like saying cars use gasoline for their movement, whilst ignoring the incredible mechanisms that make cars go. Motor proteins are just another validation of *Blume's Law Concerning the Origin of Living Cells*.[8,9]

Many biologists will cite what are called *morphogens* when trying to answer the question of what entity directs the morphing of a zygote into a fully formed infant.

Morphogens are diffusible biochemical molecules that supposedly determine cell type during embryonic development. A basic definition of a morphogen is:

... a substance governing the pattern of tissue development in the process of morphogenesis, and the positions of the various specialized cell types within a tissue. More precisely, a morphogen is a signaling molecule that acts directly on cells to produce specific cellular responses depending on its local concentration. Since morphogens diffuse through the tissues of an embryo during early development, concentration gradients are set up. These gradients drive the process of differentiation of unspecialised (stem) cells into different cell types, ultimately forming all the tissues and organs of the body.

Few morphogens have been found when nearly an infinite number of morphogens are needed. But they are already credited with being capable of forming all of our different kinds of cells, and, the entire human body. The writer of the above quote credits morphogens with cell differentiation, the formation of all of our different kinds of cells, from a blastocyst whose cells are all identical. It's a lot like saying DNA holds the plans for our entire bodies, when there isn't nearly enough coding in DNA to make even one small body part. The writer of the above definition has no idea how morphogens determine and control cell type and the formation of an embryo.

One paper I utilized discussed how morphogens "signal", "promote", "repress", "activate", "inactivate", "determine", "decide"… It's so easy to credit a little known and complex biochemical as the director of cell type and embryonic formation. Talk is cheap. No matter what kind of biochemical is credited as a cell type determiner, and is added to the list of cellular biochemicals, the biggest problem remains, **how** any molecule is able to move around inside and outside of cells. What directs their movement, and what is their motor system. What actually directs cell type and embryo formation? If the few morphogens that have been found truly do control cell type and the development of the embryo, what mechanism does it use to do this? Morphogens in a non-living entity are useless molecules. There aren't enough specific types needed to form a pinky finger, much less all body parts anyway. If there were, a control center must drive and control the morphogens themselves. Using morphogens to try to explain how cells and embryos are controlled and directed just adds another layer of control problems for biologists to explain. A technical and complex paper on the subject had this conclusion:

In the future, progress will derive from *a similar type of physical, theoretical, and experimental approach at the cellular and subcellular levels: How morphogens and their receptors are moving inside cells and at the extracellular matrix.*[10]

This writer does recognize the need for internal and external cellular control, which I will discuss in the next chapter. He also states that scientists don't know now, but someday, in the future they may. Another paper said:

However, even for those morphogens that have been identified, **we still do not understand crucial issues such as how morphogens are moved through a tissue, how a gradient is maintained, and how morphogens coordinate growth and**

patterning… *Advances in similar techniques **will reveal more morphogens** in vertebrates and will allow us to determine how their gradients are regulated.*[11]

Again, someday in the future they will learn more, but they certainly don't know now. In other words, figuring *The Puzzle* out is left for scientists of the future, because the present ones sure can't figure it out. At this point in time we humans are stuck with illusions doled out by scientists on a regular basis. One of my favorites and one of the most exciting is that if we can only dig up the DNA of a dinosaur, we will be able to rebuild it, since we would have its entire body plan. We will actually be able to make a living-breathing dinosaur exactly like one that lived hundreds of millions of years ago! Oh my god, this is a thrilling illusion. Why, we are only a few decades away from doing this incredible feat. After reading this chapter, do you think this illusion is valid? Or is it a second illusion concocted to support a first illusion? An entire series of movies were made about that very subject. The first was *Jurassic Park*. It was a wonderful movie, in that the digital recreations in that movie were astounding. And they probably are the closest we will ever get to recreating dinosaurs and seeing what they might have been like in real life.[12,13]

One of my favorite illusions is one called *homeobox genes*. These genes are widely advertised as the ones that make our body parts during embryonic development. The homeobox gene family contains an estimated two hundred and thirty five functional genes that can only act as a tool in protein synthesis. Sixty-five are pseudo-genes that are non-functional. But in this illusion, homeobox genes supposedly direct cellular differentiation during fetal development, even though genes have no capability of doing so. Mutations in these genes are responsible for a variety of awful developmental disorders. For example, mutations in the HOX group of homeobox genes typically cause limb malformations. Changes in PAX homeobox genes often result in eye disorders, and changes in MSX homeobox genes cause abnormal head, face, and tooth development. Additionally, increased or decreased activity of certain homeobox genes has been associated with several forms of cancer later in life. The list of genetic illnesses and disasters associated with homeobox genes is long. The list of positive attributes and body parts that homeobox genes can make is non-existent. Like all genes, their only capability is to hold code for proteins. And these are the genes that, when they mutate and are naturally selected, supposedly are responsible for making all parts of the human body, and the bodies of all species. Mutations in these genes are 100% disastrous. Basic math shows these genes cannot code for any body parts. But the illusion continues that they do.[14,15]

Actually, the Genome Project and Dolly shouldn't have been the only nuclear bombs to the notion that DNA in the form of our genes controls cell function and holds the plans for the entire human body. Scientists have known for decades that the amount of information required to form the human body would almost infinitely overwhelm any known biological entity. The proof of this and the math required to figure it out is basic. Evolution hijacked DNA almost the moment Watson and Crick announced the discovery of its structure. DNA became evolution's *raison d'être*. The illusion of modern evolution, and its entire foundation, random mutations, was born. The source for the gradual changes that evolution needed to explain itself had been found! The immense question here is why would respected sciences such as genetics and biochemistry allow themselves to be hijacked by evo-illusionists? Why don't geneticists and biochemists speak out? Why do they, as a whole, accept evo-illusions, and even help to support and promote them? Biochemists, geneticists, and scientists from other related scientific fields need to take a real good look at themselves and ask these questions. Will they? The answer, of course, is "no". If the notions that naturally selected random genetic mutations form species and all organs and body parts collapse, and that every cell in every organism contains DNA that holds its entire body plans and control center collapse, the entire science of evolution collapses. These illusions must be preserved and promoted at all costs.

Chapter 8

Are We Just a Bunch of Mistakes?

It has become appallingly obvious that our technology has exceeded our humanity. -Albert Einstein

Have you ever wondered what the fate of humanity on Earth will be? How long can we live on this planet? Are we humans evolving to be better and better as the centuries, and millennia go by? Will our descendants all be intelligent beyond our wildest imaginations due to evolutionary improvements? Will they be more skilled at every field of endeavor? If you go by what the biological sciences have to say about mutations, the fate of humanity can be predicted; and it doesn't look good. In fact, it looks nightmarish. To give you an idea why that's the case, I would like to cite a paper written by Leslie A. Pray, Ph.D. titled *DNA Replication and Causes of Mutation*, which explains the foundation of evolution. If Dr. Pray is correct, the future of humanity is bleak. My feeling is that when Dr. Pray wrote this, she naively wasn't considering the ramifications of evolution, and the science that goes along with it, for humanity, which I have found to be the case with virtually every evolution writer. Dr. Pray wrote:

While most DNA replicates with fairly high fidelity, mistakes do happen, with polymerase enzymes sometimes inserting the wrong nucleotide or too many or too few nucleotides into a sequence. Fortunately, most of these mistakes are fixed through various DNA repair processes. **Repair enzymes recognize structural imperfections between improperly paired nucleotides, cutting out the wrong ones and putting the right ones in their place.** *But some replication errors make it past these mechanisms, thus becoming permanent mutations. These altered nucleotide sequences can then be passed down from one cellular generation to the next, and if they occur in cells that give rise to gametes, they can even be transmitted to subsequent organismal generations. Moreover, when the genes for the DNA repair enzymes themselves become mutated, mistakes begin accumulating at a much higher rate. In eukaryotes, such mutations can lead to cancer... If DNA repair were perfect and no mutations ever accumulated, there would be no genetic variation—and this variation serves as the raw material for evolution. Successful organisms have thus evolved the means to repair their DNA efficiently but not too efficiently, leaving just enough genetic variability for evolution to continue.*

I highlighted the sentence about repair enzymes because I want to know what on earth Dr. Pray thinks does the "recognizing" and "repairing"? She just states the "repair enzymes recognize imperfections... and cut out the wrong ones". Does she have any idea what a massive amount of intelligence this sequence of events would take, and the amount of movement and transposition of molecules as well? Finding and fixing errors on any entity composed of 3.2 billion parts would be an impossible task for a human without a computer and specialized machinery. But inside of cells so small that 10,000 fit on the head of a pin, non-living molecules can do the job just fine. Amazingly, Dr. Pray isn't even curious about how this is accomplished. She discusses it like it's a ho-hum event. She perfectly demonstrates one of my major points of this book. What selects and directs the movement of molecules inside of a cell? How are the molecules able to move from one location to another, just the right ones in every case?

Dr. Pray notes that mutations can lead to cancer. Articles like this showing how genetic mutations form all species, organs, and biological systems usually give example diagrams of copy errors in the code, something like the example I gave in the second chapter: ...**AG**TACGCTT... makes a copy error that results in: ...**GA**TACGCTT... Again, notice the G and A are transposed. As the paper above states, mutations, the heart and the entire basis of evolution, occurs when DNA replicates during a process called *meiosis,* which is cell division during the formation new gametes. (sperm and egg) To see a short video of what occurs during meiosis, go to my website and watch video number 6, *Meiosis: The Forming of Gametes* at : www.thednadelusion.com (Alternative: www.thednadelusionblog.wordpress.com). When a mutant sperm fertilizes an egg, or a mutant egg is fertilized, these copy errors are then passed on to the next generation. These mutations build up over eons. According to evolution, they produce the formation of all new species, biological systems, and body parts.

To give you an idea of why random DNA copy errors cannot produce functional biological systems or new species, even if they have the ability to do so, imagine having a text in your computer, such as the Gettysburg Address. You are going to copy and paste the text to a blank sheet in your Word program. You are then going to copy the first paste, to a second blank sheet, and repeat this scenario one thousand times. You also have a program that makes several random text errors every time you make a copy. After making your thousand copies, what would the Gettysburg Address look like? The speech would not improve a lick. Of course it would be completely degraded. It would become unusable mush. The same is also true with DNA coding. Random copy errors, or mutations, would turn DNA coding into mush. Because the coding in DNA is unrecognizable and invisible to virtually all humans, it's easy for evo-illusionists to convince people that beneficial mutations can occur that make new species and biological systems. If evo-illusionists had to convince people that random changes improved recognizable text such as the Gettysburg Address, no one would go for it. But convincing people that ...**AG**TACGCTT... is better than ...**GA**TACGCTT... and mistakes like these formed all species including humans, our brains, eyes, limbs, and all organs, is easy. Who can argue? I had a contractor build my house in 1978. I will never forget that he told me that the less people know, the easier they are to fool, and the more they pay. That notion works in so many venues, and particularly in this one: evolution.

Dr. Pray, in her quote above, doesn't realize she is describing a doomsday scenario for humanity. According to evo-illusion, if deleterious mutations sneak through the biochemical correcting processes, natural selection eliminates them in the wild. The animals with the deleterious mutations are killed due to predator/prey relationships, changes in climate, or competition for food or mates. The mutations that survived DNA's correction process, and also survived natural selection, are responsible for the formation of all living species, all organs, and all biological systems that have ever existed, and that exist today. Yes, even your brain, intelligence, consciousness, liver, heart, blood, skin... everything is formed because of these incredibly rare copy errors that are "selected for", according to evolution. But evolution is a science with tunnel vision. It's meant for animals in the wild, but not plants and people. Strong plants do not kill and consume weaker ones, and they don't do battle for mates. The major factors connected to evolution's selection and correction processes are mostly excluded for plants. Plants are eliminated from most of the effects of natural selection, which should make one wonder if evolution is capable of originating and forming all of the millions of plant species on Earth, and their vast array of characteristics and biochemical processes. Do floods and droughts create enough selection to form flowers, weeds, artichokes, and trees, and all of their complex biological devices? Are evolution's illusionists distracting their audiences by focusing almost completely on animal evolution? It's a pretty good trick, and one used by stage illusionists all the time. From the looks of it, evo-illusionists use this trick as well. "Get them to focus on animals, so they don't think about plants." Almost all discussion about evolution centers on animals. Plants are rarely discussed. And certainly the foundation of evolution, the genetic system, is never discussed. There is no imaginable way that DNA, mRNA, tRNA, ribosomes, the incredible protein-forming machine shops that each cell has, and the dozens of enzymes that push the process of protein formation, could possibly come into existence by evolutionary processes; by random mutations and natural selection of an incredible process that did not even exist when the Earth first formed. Could a mindless process invent DNA synthesis and proteins when none existed before? As I say, the biggest problem for evolution is not design; it's invention. The notion that evolution could invent and develop the genetic system is ludicrous. Santa Clause is a more believable notion.

Regarding the evolutionary outlook for humans: humanity has been eliminated from most natural selection activity by our own intelligence and ingenuity. Natural selection improves the health of animal populations; but not human populations. The vast majority of mutations are deleterious for humans. The fact that bad mutations that occur during the formation of human sperms and eggs aren't completely corrected, as cited by Dr. Pray, and are not removed by natural selection, is a potential catastrophe for humanity. Humans no longer exist within the predator/prey system credited by evolution's illusionists with saving beneficial mutations and removing the deleterious ones. Humans are no longer animals in the wild. Since humans left the predator/prey system thousands of years ago, there is no entity that removes harmful genetic mutations from the human population. Weaker humans can procreate nearly as frequently as do the strong healthy ones. Humans have found ways to eliminate practically all of the environmental problems that might select the strong and good mutations, and eliminate the bad. The vast majority of humans have efficient shelter and protection from most weather related calamities. They have incredible food resources, and a plentiful supply of mates. We don't have to do battle to the death or cause severe injury to our mating opponents to win our mating partners. We don't have to fight for food. Humans have been nearly completely eliminated from the natural selection process that drives evolution. If evolution is valid, the net result is we are gradually being poisoned by our own genetic system. Mankind will become inundated with mutants like the one in Fig. 8-1. At least 98 percent of mutations are deleterious or neutral. The mutations that escape cellular biochemical correction will remain in the human population, and be added to, generation-by-generation. The future of mankind is that everyone will have cancer and severe physical limitations and illnesses due to this ongoing buildup of bad mutations. If modern biology is correct, the human population on Earth will be destroyed by its own genetic system. The only hope for humanity is if there is a currently unknown or unrecognized natural controlling and correcting mechanism that is able to repair 100% of DNA replication errors, and that works above the cellular level correcting systems. Hopefully, there is such a controlling mechanism, and it can correct the massive numbers of mutations that have formed and will be forming in human DNA. If there is no correcting mechanism that can take over for humanity's loss of most natural selection processes, we humans will inhabit the Earth for a comparatively short time. We will have our own consciousness and intelligence, and our ability to protect ourselves from predation and most environmental upheavals to blame for our demise. Is there an entity that is in a hierarchy above DNA that could save humanity from total collapse?

Fig. 8-1

When gametes make copies of themselves, DNA replication typically produces only one error per 1,000,000,000 (one billion) base pair copies. That is an astounding fact. Mankind isn't remotely close to being that accurate at doing *anything*. The processes of cell and gamete replication are unimaginably accurate, far beyond any human's ability to grasp. It's easy to write about the accuracy of genetic copying, but to actually think about it, and to try to visualize it is beyond comprehension.

During cell division in bacteria, called *binary fission*, base pairs are copied at the rate of about 1,000 per second. At that unimaginable rate, the copying is also inconceivably accurate. About one error is made per billion base pair copies also. To give you an idea of how accurate this is, consider *Escherichia coli* bacteria. It has 5 million base pairs in its DNA that need copying with each cell division. At the rate of one error in one billion copies, E. coli makes about only one base-pair error per eight generations. DNApolymerase and template DNA correct most of these. Can you imagine how smart DNApolymerase must be? I wonder if they go through some kind of training regimen, or schooling, or something. How do they know which base pairs are incorrectly placed, and which bases are needed to complete the repair? The errors that sneak through correction processes remain in the genome permanently.[1]

Scientific illusionists promote that all modern life originated from cells called prokaryotes (bacteria). Is it imaginable that the mind-bogglingly accurate gene copying of bacteria could produce enough errors that would invent and develop all complex biological systems of all living organisms? Could a single bacterium and its offspring, with so few copy errors, if left alone for billions of years, eventually produce humans? Evolution says that's the case. Are we humans a product of genetic copy errors starting with bacteria that are unimaginably rare? In a system that only makes proteins? That occur only every 8 generations?[1]

Dr. Pray wrote a paper on the subject of bacterial mutations. She discusses getting a cut, then touching a doorknob with one bacterial cell on it that gets into your cut:

Even if only a single S. aureus cell were to make its way into your wound, it would take only 10 generations for that single cell to grow into a colony of more than 1,000 ($2^{10} = 1,024$), and just 10 more generations for it to erupt into a colony of more than 1 million ($2^{20} = 1,048,576$). For a bacterium that divides about every half hour (which is how quickly S. aureus can grow in optimal conditions), that is a lot of bacteria in less than 12 hours. S. aureus has about 2.8 million nucleotide base pairs in its genome. At a rate of, say, 10^{-10} (one in ten billion) mutations per nucleotide base, that amounts to nearly 300 mutations in that population of bacteria within 10 hours! To better understand the impact of this situation, think of it this way: With a genome size of 2.8×10^6 (2.8 million) and a mutation rate of 1 mutation per 10^{10} (ten billion) base pairs, it would take a single bacterium 30 hours to grow into a population in which every single base pair in the genome will have mutated not once, but 30 times! Thus, any individual mutation that could theoretically occur in the bacteria will have occurred somewhere in that population—in just over a day.[2]

Dr. Pray gives bacteria a one in ten billion rate of mutations, ten times more rare than my research showed. So her entire example population of bacteria, which grew from just one bacterium to one million in 10 hours, would have 300 mutations. Bacteria like *S. aureus* have been around for 3.5 billion years. At the rate of 300 mutations per 10 hours, if you could follow a single *S. aurueus* or the like growing into one million repeatedly over and over for 3.5 billion years, the final population of S. aureus would have had 3×10^{13} (30 trillion) mutations. This means that the S. aureus bacteria have had 100% of its genome changed 11 million times since its beginnings on Earth. And guess what. They are still bacteria. Is the correction function of S. aureus DNA efficient enough to correct all of these errors? If only .05% of errors snuck through correction processes, that would still mean S. aureus bacteria, using Dr. Pray's example, would have changed 100% of its genome 5,500 times. If evolution were valid science, wouldn't this immense number of mutations be enough to cause S. aurueus to evolve into something other than bacteria? Like maybe a dinosaur, or a human, or a banana? Or some other kind of species? Or complete mush? But sadly for Dr. Pray, and evolution, they are still bacteria; and bacteria are the most numerous, stable, and successful group of living species in the history of the Earth. If you considered the number of mutations that would occur in developing the modern population of S. aureus, using Dr. Pray's figures, the there would be an almost infinite number of mutations. Certainly enough to evolve the most complex of multicellular organisms, if that's indeed how they arose. Dr. Pray innocently thinks she is proving evolution. She is actually proving the absurdity of the theory.

Since humans have 3.2 billion base pairs in their DNA, they make, on average, about three or four errors with every cell division. But since almost 98 percent of the DNA in the human genome doesn't code for proteins or any known entity, 98 percent of the copy errors in human cells should be insignificant and can be dismissed as far as evolution goes. And since 98 percent of copy errors are harmful or neutral, that leaves 2 percent of 2 percent or four one hundredths of one percent of mutations for evo-illusionists to use as their explanation for evolution. No scientific illusionist has been able to cite any mutations that make healthy utilitarian tissues beneficial to any species population anyway.

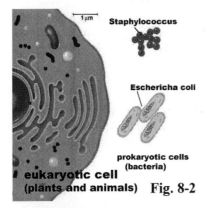

Fig. 8-2

The two most common types of cells are bacterial cells called *prokaryotes*, and the cells of all plants and animals called *eukaryotes*. Prokaryotes have a nucleus that isn't contained in a membrane. Their nucleus is diffusely spread throughout the cell and is difficult to view in a microscope. Notice how the chromosomes in Escherichia coli in Fig. 8-2 are dispersed. Prokaryotes have far fewer internal organelles than do eukaryotes. On the other hand, the nuclei of eukaryotes are tightly confined inside of a nuclear membrane, and they are easily visible under a microscope. It's interesting to note that there are no bacteria in the process of evolving a nuclear membrane on their way to becoming eukaryotes, sadly for the theory of evolution. In fact there is no imaginable way that could happen. Try to think up any imaginary steps between the loose chromosomes, which diffusely float around inside of a bacterial cell, and the nucleus of a eukaryote that is tightly bound by its Saran Wrap-like nuclear membrane. There are none. If a eukaryote were the size of a basketball, a prokaryote would be about the size of a Ping-Pong ball. Size-wise, there are no intermediate cells that exist between prokaryotes and eukaryotes. If evolution were valid, there certainly should be. Further killing the illusion that prokaryotes evolved into eukaryotic cells as evo-illusionists say they did, and then eventually into humans, is the fact that bacteria would have had to add 3.195 billion base pairs to its DNA over 500 million years to reach the 3.2 billion base pairs we humans have. This is a 640-fold increase! There has never been an instance where scientists have found examples of any species increasing the size of its genome. So the notion that bacteria evolved into humans can be logged as just what it is: a bad and amateurish fable believed by many.

The illusion has worked well for evolution for decades, ever since Watson and Crick determined the molecular structure and functioning of DNA in 1953. But, piece-by-piece, DNA itself has destroyed this illusion. Dolly revealed that DNA doesn't control cell and body type. The Genome Project showed that there are barely enough base pairs in our DNA to construct the 90,000 to 2,000,000 proteins needed by the human body. There certainly aren't enough to control the billions of molecules inside of cells, hold the plans for the human body and all of its cell types, and control all cellular functions that need controlling inside and outside of cells. The foundation of the illusion of evolution has completely crashed. This fact has been kept quiet; few supporters of evolution have any realization. The scientific illusion of evolution lives on, even though its foundation has been completely destroyed.

Cells are tiny machines, made up of proteins and other biochemicals, instead of plastic and metal, as are man-made machines. Proteins make up 20 percent of the weight of cells. Proteins have individual shapes according to their length, order, and the type of amino acids that make up their chains. Both ends of proteins are negatively charged (-) which makes them stretch out because the ends repel each other, similar to two magnets with the same positive (+) or negative (-) ends meeting up. When a signaling molecule comes along, such as a hormone, the signaling molecule has a plus (+) charge, which is greater than the negative charge on the end amino acid in the protein. If the signaling molecule attaches to the end amino acid of a protein, the ends of the protein will now attract each other, causing the protein to change shape. If the signaling molecule breaks away, the protein will then go back to its original more stretched out shape, because the negative ends will again repel each other. The shape of a protein molecule is determined by what its function is. Their shape allows proteins to fit other molecules they are affecting like miniature three-dimensional jigsaw puzzle pieces.[3,4]

The manufacturing apparatus that makes proteins is itself made of proteins. To see what an immense problem this is for evo-illusion, imagine if there were one single brick manufacturing plant in the world, and all bricks that ever existed had to be manufactured by this plant. The plant is completely composed of bricks. The question should of course arise, where did the bricks come from that were utilized to construct the brick-manufacturing plant? Bricks made by the plant are going to be used to build a building. The plans for the entire building are wrapped up in a tiny blueprint inside a hollow core in each brick. Is there any way these plans could direct the construction of the building? No matter what mechanism might be used to locate and utilize the plans, the plans could not aid in the construction of the building unless they were dug out by an intelligent engineer capable of reading and acting on the plans; in other words, an external control source.

The same is true with the notion that the DNA held inside of all cells in a human holds all of the plans for the human body. Even if that were the case, there is no known controlling and communicating feature of DNA that would give it the ability to form a developing fetus. I find it amazing that the supposed *source* of evolution's driving mechanism, DNA, and the apparatus needed for protein synthesis, is itself unexplainable by the processes of evolution. Evolution could not occur until DNA replication and cell division began, which means DNA and protein synthesis must have existed before evolution existed. So what was the source of DNA and its functions? Why doesn't this fact alone kill the illusion of evolution? The foundation for evolution, copy errors in the machinery that performs protein synthesis, has no possible evolutionary source. All of the parts would have to exist together for each part to have utility. Parts of a system that made proteins could not have originated,

AKA been invented by, mutations in that non-existent system itself. For some strange reason no evolution scientist I know of even considers this to be a problem or seems to even wonder about it. I do.

Chapter 9

The Inner and Outer Domains Further Destroy the Illusion of DNA

Organic chemistry is the chemistry of carbon compounds. Biochemistry is the study of carbon compounds that crawl. — Mike Adams

Modern science has discovered an incredible amount of scientific information in the last 150 years. Each new discovery begs a new question about its origination and control. The scientific illusion that covers this fact must grow exponentially with each new discovery, or the entire illusion will crash. [6]

So what is the control center of living organisms? There are two major domains that need to be addressed when considering cellular control centers:

(1) The *internal cellular domain*: With all unicellular and multicellular organisms, the control of the various molecules, molecular machinery, and the organization of the domain *inside* of each cell.

(2) The *external cellular domain*: With all multicellular organisms, the control of the various entities and the organization of the overall multi-cellular organism itself, its embryonic development, and events that occur in the domain *outside* of its cells. Deals with the interactions of individual cells, cell groups, and their molecules.

These are two very different domains with two very different sets of needs and problems. Are the control centers for these two domains one and the same? Or are there two entirely different control centers for each domain? It's so easy to say that the nucleus of the cell with its DNA controls everything. That has been an accepted scientific illusion for decades. "It's in the genes," is such a commonly used phrase. But when you look at the facts, it just isn't possible. Dolly demonstrated that DNA isn't even capable of controlling cell type. That being the case, it certainly can't control the design and embryonic formation of long bones, brains, blood vessels... Nor can it control reactions to cellular macro-events that happen to multicellular organisms. These are events that take place outside of cells in the external cellular domain.

To give you an idea of what I mean by events that occur outside of a cell versus those that occur inside, a cut made by a sharp razor is a great example. Imagine what transpires if you cut your face with a sharp razor blade. Since cells are so small, remember 10,000 skin cells would fit on the head of a pin, it's unlikely the blade itself would actually damage any of them. Cells would be separated, and some may be torn apart by the cut, but cells couldn't actually be cut. A razor blade is about 100 microns (1/10,000 of a meter) thick. The diameter of a skin cell is about one-ninth the thickness of a razor blade, or about 11 microns (1/90,000 of a meter) thick. Razor blades are way too thick to cut any cell membrane (outer wall) of a skin cell. Intact cells wouldn't have an awareness of the cut unless they have some sort of consciousness in their external cellular domain. Fig. 9-1 shows skin cells as viewed under a microscope. The edge of a razor blade would be almost as wide as the photo.

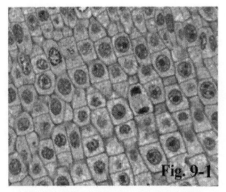
Fig. 9-1

To see a short video on wound healing go to video 7, *Steps to Wound Healing* at www.thednadelusion.com (Alternative: www.thednadelusionblog.wordpress.com) Notice how each cell involved has to know exactly where it is supposed to go and what its task is. When cuts, such as the razor cut, occur in our skin, an immense healing process is undertaken. Tens of millions of various types of cells must act in concert. These cells have to be conscious of what has taken place in the external cellular domain. *Platelets,* blood cells that trigger *hemostasis* or clotting, rush into the wound area. They initiate the release of *platelet-derived growth factors* into the wound that causes the migration and division of cells during the *proliferative* phase of healing. How do the platelets "know" what occurred in the external cellular domain outside of their cell membranes? Can they see? Can each platelet think and react? Blood clotting alone has its own complex biochemical pathway. Next, *inflammation* begins. Bacteria and cell debris are consumed, a process called *phagocytosis*. Foreign fragments are removed from the wound by white blood cells. The *proliferative* phase is characterized by growth of new blood vessels into the wound, the laying down of fibers called *collagen*, which produces scar tissue, contraction of the wound, and new skin formation. The final major stage is *remodeling*. During this stage, the tissues that heal the wound mature and become 80 percent as strong as were the original tissues before being cut. The actual course of events is far more complex and impressive than this short paragraph depicts.[1]

All steps require cells to have *awareness* about what happened, or needs to happen, in the domain *outside* of their cell membranes as well as inside. In the external cellular domain of cells, tens of millions of various *cells* must be guided to the wound site so that healing can take place. In the internal cellular domain of each cell, tens of millions of molecules must be produced, then sent through the cell membrane (wall) and guided to their needed job site. So you see, there are two very different domains, with two very different requirements, and two very different functions. What happens in the external cellular domain to the millions of cells that take part in the healing process alters what goes on in the internal cellular domain of each of those millions of cells. The internal and external cellular domains are inexorably connected. Are they controlled by the same entity? If not, they must be in very intimate communication.

How do modern scientific illusionists explain the entity that informs and controls cells that rush in to repair injuries? This is what the *DNA Learning Center* has to say in their video demonstrating wound repair:

If you are hurt, your cells work together to repair the damage. They communicate using their own language of chemical signals.[2]

They communicate using "their own language"? And by "chemical signals"? That's it? See how scientific illusions are so easy to build? The DNA Learning Center exists to teach students. They do a wonderful job of explaining many biological functions, and they have many fascinating scientific videos on DNA and wound healing and many other subjects. Why don't they simply relay the information that no scientist has any idea what the control center is that coordinates the movement, communications, and functions of cells whose job it is to repair wounds? Why do they need to make something up to cover and explain an unknown? Isn't the coordination, organization, and obvious need for a controlling entity of the cells and molecules in and around a wound one of the most extraordinary events in all of nature? Why is it just glossed over as if it's a non-existent or unimportant part of wound healing?

It could be argued by someone who is not familiar with biochemistry, and the internal functioning of cells, that the brain is the control center of wound healing. The argument could go like this: *if a cut occurs, nerves send a signal to the brain, which in turn directs cellular and biochemical responses.* The problem here is that the brain, and all of the neurons that make up the brain, and all nerves, need all of their biochemical molecules to be controlled, coordinated, and directed by a control center, just like all of the repair cells that travel to the wound site. Trying to support the brain as the control center only exacerbates the problem. The brain as control center only presents more problems. The controlled actions of hundreds of billions of more molecules and millions of cells need further explaining. What controls, coordinates, and communicates with all of the molecules in the cells of the brain, and all of the molecules inside of the cells that make up nerve connections?

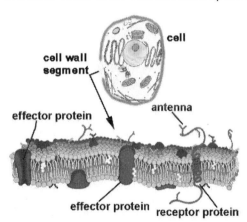

Fig. 9-2 Section of a Cell Membrane

DNA cannot reach out and control other molecules swimming around the inside and outside of the cell. There is something above DNA that is far more impressive, powerful, and more conscious than DNA in the hierarchy of possible control centers for both the internal cellular domain and external cellular domain of cells. My cut-skin example also begs a deeper puzzle: the growth and development of a fetus requires an unimaginably complex and powerful control center for the two domains. What takes place in the inside domain of each cell of a maturing fetus, and what takes place on the outside domain are two entirely different spheres of influence that require direction and control.

Cell membranes (Fig. 9-2), the outer capsule of animal cells, separate the internal cellular domain from the external cellular domain. It allows only certain molecules into the cell, and it relays messages via a chain of molecular events. There are thousands of protein molecules in the cell membrane designed to pass information in the form of molecules from the external cellular domain to the internal cellular domain. These proteins can be divided into two groups: *input proteins*, or *receptor proteins*, and *output proteins* or *effector proteins*. Receptor and effector proteins are much like our eyes, nose, ears, and taste buds. Information that forms on the outside of cells is transferred through the cell membrane to the inside of the cell through these two specialized types of proteins. Receptor proteins have a kind of antenna that extrudes out, that can accept information in the form of specialized molecules from the outside of the cell. Information can be transported by toxins, hormones, sugars, histamines... Incredibly, receptor proteins with their extruding antennae actually migrate around the cell membrane searching out stimuli and information. For every entity the cell can respond to, there is a different antenna; a different receptor protein. Receptor proteins are designed to be specific to the type of information or molecule they can receive. Effector proteins also migrate around the cell membrane and search out receptor proteins that have gathered information necessary to keep the cell functioning. Receptor proteins pass molecular information on to effector proteins, which extrude on the inside of the cell membrane. The effector proteins then pass the information on to a third type of molecule specialized to "swim" to the location inside of the cell that can use the information gleaned from the receptor and effector proteins in the membrane, so that information can be acted upon.[3,4]

The entity that directs the goings on inside of cells, and the goings on outside, is mysterious beyond belief. It would seem that as far as science has come in understanding cells, we would have some notion about what the *Great Director* is. But we have no idea; not even an inkling of and idea. We have some bad notions, like morphogens, that don't come close to answering the great questions cellular domains pose; but no good ones. This *Puzzle* is far greater than the puzzle posed by DNA before 1953 when Watson and Crick did the solving. Will it ever be solved? As smart as modern science has become, it's not good enough. We have no idea which means we probably never will. I wish I could live long enough to see the solution to this puzzle; and the puzzle of the existence of the universe, and the source of life. But sadly for me, I have no chance. At least I've lived when some really great mysteries were solved. Actually I should be satisfied with the solution to the mystery of what those tiny water balloons that make up all living things are. I saw the advent of the computer, and man land on the moon; and cars and airplanes and indoor plumbing... So I guess I can't complain. But I want to know *everything*!

Chapter 10

So Then What Makes Babies?

There is really nothing quite so sweet as tiny little baby feet.- author unknown

What does modern biological science have to say about the formation of the vast number of different cells in the human body, and what controls the formation of those cells in the womb? What controls and guides the embryonic formation of the entire human body? To get an answer, I chose a scientific educational organization, *Stated Clearly*, to get a good overview of what science has to say about what controls the formation of a zygote and its development into a fully formed infant. Stated Clearly's passion is described on the lead page of their website:

We are in love with the scientific process and the art of critical thinking. We feel that an enhanced ability to think critically and ask probing questions will enable people to become better scientists and also aid them in every aspect of life from work and politics to family life and personal relationships.[1]

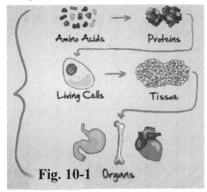

Fig. 10-1 Organs

I wholeheartedly agree with their statement of purpose. But do they follow it? Stated Clearly clearly states that they are an honest scientific organization. They say that when writing an article or the script for an animated documentary, they start by consulting as many textbooks and scientific journals as they can. When possible, they contact researchers and authors of specific papers to clarify any of their questions. In doing this, they've made many contacts in the scientific community. They now have a group of researchers who review their content for factual accuracy before it's published. They list twelve Ph.D. biologists on their panel of advisors. NASA, Emory University, Georgia Tech University, and the National Science Foundation back them. In other words, they are the perfect group to answer my challenge. Here is what Stated Clearly has to say about human development in their documentary video, *How DNA Works*. Fig. 10-1 is a diagram made to help the viewer understand DNA, produced by Stated Clearly:

Amino acids make up proteins. Proteins, along with other chemicals, make up cells. Cells make up tissues. Tissues make up organs, and organs, when they are all put together and functioning, of course, make up creatures like you and me. These proteins that make up our bodies, and keep in mind there are millions of different kinds of proteins, they each have to be formed in the perfect shape in order to function. If they're the wrong shape, they usually won't work. That's where DNA comes in. DNA does a lot of interesting things, some of which we don't fully understand. But one of the functions is to tell amino acids how to line up and form themselves into perfect protein shapes. In theory, if proteins are built in the right time and in the right place, everything, from cells, to organs, to creatures, will come out just fine.[2]

"Everything will come out just fine?" Is that "stated clearly"? The gaps in their "stated clearly" science are immense. In introducing and then naively ignoring these immense gaps, they also create a scientific illusion: the illusion that scientists know how humans and all of their incredibly complex parts actually form embryonically and by evolution. In reality, they are in a scientific fog. They either don't know that they don't know, or they don't want their audiences to know that they don't know. In either case, Stated Clearly presents a scientific illusion. They certainly don't follow their stated purpose.

The first step in their above chart is that amino acids make up proteins. Fine. Scientists know **what** DNA and friends do to produce protein molecules, but they have no idea **how** they do it or **what** controls the process, which seems to me to be the biggest puzzle of all. But amino acids do make up proteins through protein synthesis. The biggest gap here is that amino acids only make proteins *inside* of cells. Cells are the third step on this Stated Clearly diagram when they need to be the first. Not one protein molecule exists on Earth, and probably in the universe unless the molecular equipment inside of a cell made it. Not one. So Stated Clearly's third step, proteins make living cells, is a complete scientific illusion. Proteins make up only 20 percent of living cells. Eighty percent of the molecules in living cells are not proteins. What is the control center in the cell that makes up all of these biochemicals? What organizes them? What entity in a cell communicates with biochemicals, and determines what kinds of cells need to be constructed? What notifies them where they are needed? What originally caused these cells to come to life? What guides cell type? This gap is immense but ignored by Stated Clearly. So is the next gap: the formation of tissues from cells. The embryonic cells that start the formation of a fully formed human infant differentiate into over 400 different types of cells. What determines how many of each of the 400 cell types needs to be produced? What controls where they go in the body? What guides this whole process? What causes those tissues to form the organs that Stated Clearly says is the last step on its diagram? The diagram is nothing but an illusion, but many people will see it and be comfortable that it's accurate, and that we humans know how proteins, cells, and our bodies formed. This is a perfect example of the more we know, the farther we must realize that we are from figuring out the puzzles and dilemmas posed by living nature. But the larger grows the illusion that makes the audience think we have solved nature's mysteries involving the development of cells and the human body. In reality, all we can do is observe, marvel, and study processes that we have absolutely no idea how they occur We are much like the ancients who had no idea how lightning and thunder occurred.

By far the most daunting challenge for modern science is to determine *how* each of the billions to trillions of cells in a developing fetus achieves their cell type, shape, and position so they can perform their individual functions. *How* these billions of cells are guided to their locations in the forming fetus in perfect order is always credited to our genes, an entity that has absolutely no capability of doing that job. Fig. 10-2 shows the development of an embryo during its first 23 days. Take a look at Day

12, and try to imagine how the tiny dot, the nucleus, inside of any of those cells might control the forming and shaping of the embryo during days 13 through 23. It's taken for granted that the genes are the control center for embryonic development. Few people question. It's similar to the "given" in a geometry problem, or the way the ancients accepted that lightning was created by gods such as Thor and Zeus. Few ancients questioned Thor and Zeus, and few question the DNA illusion today. An embryo needs an overall control center that controls what occurs outside of the group of cells that are developing, as well as what occurs inside those cells. A control center inside of a cell cannot possibly determine and control the formation of long bones, teeth, auditory system, muscles, intelligence... The formation of these requires an overall control center that determines how many of each cell type will form, what directions the growth of those cells will take, what shapes and configurations the cells will have, and the position of those cells in the embryo. In other words, the entire anatomy of the developing embryo must be controlled by a control center that is responsible for reacting to *what has occurred*, in the case of the cut, and *what will occur*, in the case of the embryo.

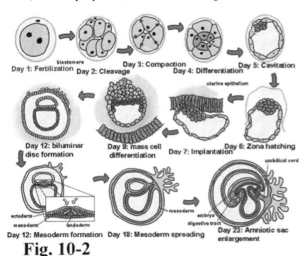

Fig. 10-2

What occurs inside of a cell cannot control the overall external cellular domain of a cell any more than my imaginary hollow brick can control its external brick domain. To see an amazing video on *The Development of a Human Embryo* go to video number 8 at www.thednadelusion.com (Alternative: www.thednadelusionblog.wordpress.com) Try to imagine any controlling entity directing the trillions of cells that you are watching develop into a human infant. What could possibly be directing such an incredible miracle that is happening right before your eyes? Certainly not the 22 million DNA codons that are hidden in the nucleus of each cell.

Take a look at Day 1 in Fig. 10-2 above again. The two dark dots are the nucleus of the original ovum, and the fertilizing sperm. There isn't room, nor is there a means, to hold the quadrillions of bits of information needed to build an entire human being. Since one single fertilized cell divides into two cells, then four, and then in nine months it becomes a fully formed human infant, where could these plans be? Common sense, Dolly, the Genome Project, basic math, and protein synthesis clearly show that the nucleus and its DNA must be rejected as a candidate for holding the plans for a complete human.

Dr. Frank T. Vertosick, Jr., the author of *The Genius Within*, has suggested that the control mechanism for cells is in the cell's cytoplasm, the liquid that supports all of the internal cell parts. It makes up the body of the cell. The cytoplasm is all that's left to credit as the control source once DNA, and all organelles and biochemical molecules are eliminated as candidates. But cytoplasm is not much more than saline solution. Salt water has no capabilities of doing anything but existing and being a good solution and support for other chemicals.[3]

A human ovum (egg) is about one quarter the size of a grain of salt. If the nucleus, cytoplasm, and cell capsule are eliminated as plan holders, is it possible that the mother somehow holds plans for her developing infant? Does an unknown entity present in the mother somehow transfer information to the fertilized ovum so that it will divide and differentiate so perfectly that it will wind up a fully developed human? If that were the case, an infant born from a surrogate mother would resemble the surrogate mother, not the ovum donor; that just doesn't happen. But is it possible *some* inherited information comes from the birth mother?

It's so difficult to give these questions the significance and weight that they deserve and that I would like to give them. Maybe bolding a few words will help:

How do all of the trillions of cells of the human body organize themselves to form, shape, and assemble all anatomical body parts, and biological systems? How do body parts and the cells that make them up form exactly where they are needed? How is the development of the shape, size, and location of each anatomical entity, made up of cells, controlled? What entity keeps an accounting of the number of cells needed for each body part during embryonic formation? What starts and stops their formation? Does each cell have a brain, eyes, swim fins, and a plan? Do their molecules? How do cells "know" what to do when they arrive at their destination?

Whew! There. That feels a bit better; but not much.

In a major news item, a team at the Hebrew University of Jerusalem announced in a paper that they've discovered the mechanism that allows embryonic cells to become differentiated cells; cells that make up the various different tissue types, such as lung tissue, liver, kidney, eye, skin... Here is a short synopsis.

Mechanism That Triggers Differentiation Of Embryo Cells Discovered Date: December 22, 2008 Source: Hebrew University of Jerusalem

Summary:

The mechanism whereby embryonic cells stop being flexible and turn into more mature cells that can develop into specific tissues has been discovered...

At a very early stage of human development, all cells of the embryo are identical, but unlike adult cells are very flexible and carry within them the potential to become any tissue type, whether it be muscle, skin, liver or brain.

This cell differentiation process begins at about the time that the embryo settles into the uterus. In terms of the inner workings of the cell, this involves two main control mechanisms. On the one hand, the genes that keep the embryo in their fully potent state are turned off, and at the same time, tissue-specific genes are turned on. By activating a certain set of genes, the embryo can make muscle cells. By turning on a different set, these same immature cells can become liver. Other gene sets are responsible for additional tissues...

They found in their experiments, using embryos from laboratory mice and cells that grow in culture, that this entire process is actually controlled by a single gene, called G9a, which itself is capable of directing a whole program of changes that involves turning off a large set of genes so that they remain locked for the entire lifetime of the organism, thereby unable to activate any further cell flexibility.[4]

I should send these people my book so they can realize that the plans for the human body don't reside in the genes. They actually think a single gene "turns off" and "turns on" and "locks" many other genes, similar to a switch, so that an embryo can form all necessary tissues to make a fully developed fetus? The notion that one gene can control what happens with, and can switch on and off a myriad of other genes, doesn't hold water. In any case, switching off and on DNA code doesn't and couldn't form body parts. DNA is only a strip of protein-making code. So again, I have questions. How does one gene communicate with all other genes? How does it pass instructions from gene to gene? How does it direct? What is its *modus operandi*? How does it know where to go to do the turning off and on? How does it get feedback that its instructions have been carried out? It's so easy to say something "switches on" and "switches off" a bunch of other genes, because humans understand switches. We use light switches and various other switches on our electronic devices every day. So the "switch notion" is acceptable and easy to understand, comprehend and believe; an easy illusion to concoct. But, again, there just isn't any *known mechanism* that causes one genetic molecule in a cell to communicate with another genetic molecule, from a distance. It certainly happens. Millions of molecules are zooming around and communicating in the inner domain of every living cell at all times. But what mechanism drives and guides them? And what allows the inner workings of embryonic cells to control what happens externally so they can control cell type and location of other embryonic and fetal cells? Which results in the formation of an entire human infant and all of its body parts and biological systems from a single cell? The biggest problem for scientists who try to support this fable is that each gene has only about 27,000 to 2,000,000 base pairs, or 9,000 to 700,000 codons. This isn't nearly enough information to make the retina of an eye, which is made up of 125,000,000 retinal cells.

The earliest known explanation for how a single cell develops into a human in only nine months involved a process called *preformation*. Preformation maintained that one of the human gametes, either ovum or sperm, holds a micro-miniature model of the adult human being called a *homunculus*. Of course, only the sperm or ovum could carry the full model human. There couldn't be half of a human in each, then they combined into one human during fertilization. So only the sperm or ovum carried the model. Of course there was a serious battle between the *ovists* and the *spermists* as to which contained the homunculus. Fig. 10-3 is a tiny person inside a sperm, as drawn by Nicolaas Hartsoeker in 1695. According to the

Fig. 10-3

theory, each full human model actually had smaller models inside, much like a series of Russian dolls that are nested one inside the other. The smaller homunculus represented earlier generations. Gee, a serious battle about a fable, just like evolution is today? You see, we still have one foot in the Middle Ages. We are no different than they were. We modern humans are programmed to make up fables about things we don't understand, just as did the ancients.[5]

Let's check out what the leading evo-illusionist in the world, and my personal favorite, Richard Dawkins, has to say about what directs the formation of the fertilized ovum into a full human infant in only nine months. If you'll recall, Dawkins wrote a NYT best selling book titled, *The God Delusion*. Of course the title of my book, *The DNA Delusion*, is a takeoff from and a challenge to his book. Dawkins thinks by pitting evolution against religious belief, and showing that the existence of a religious god is a sham, he is proving evolution. Of course he knows this isn't the case. He's too smart to not know. Dawkins outsmarts modern science and every evolution believer on the planet. What he does is called *bait and switch*. He switches his need to prove evolution with trying to disprove god and religion. It's a technique used by almost every conman, of which Dawkins certainly is one. He is unquestionably the most famous conman in modern scientific history. Like so many conmen, he has made millions with his shtick. Is he any more knowledgeable on the subject of what directs embryonic formation than the preformationists, or the Hebrew University of Jerusalem? In reality, he is just as much in the dark as every human, since no person knows. Actually Dawkins' story changes from book to book. Why not? Fables are far easier to change than is true science, because fables are nothing but figments of man's imagination. There is no scientific proof required with Dawkins' fables, so he needs no scientific evidence at all. He can't be called out for posing a phony story, because there is nothing in existence that can say he's wrong. Evo-illusion has grown so vast, with so many tales, it gets very confusing for the general public and its fans and students to discern what is real and what is made up. So everything Dawkins and his fellow evo-illusionists say is accepted as valid. In his NYT best selling book, *The Selfish Gene*, (Oxford UP, 1990) Dawkins tells his readers where the plans and directions for the formation of a human from a single cell (ovum) reside: in every cell in our bodies. Dawkins writes:

There are about a thousand million million cells making up an average human body... every one of those cells contains a complete copy of that body's DNA. **This DNA can be regarded as a set of instructions for how to make a body, written in the A, T, C, G alphabet of the nucleotides.** *It is as though, in every room of a gigantic building, there was a* **book-case containing the architect's plans** *for the entire building. The 'book-case' in a cell is called the nucleus. The architect's plans run to 46 volumes in man; the number is different in other species. The 'volumes' are called chromosomes. They are visible under a microscope as long threads, and the genes are strung out along them in order...* **The DNA instructions have been assembled by natural selection.** *DNA molecules do two important things. Firstly they replicate, that is to say they make copies of themselves... As an adult, you consist of a thousand million million cells, but* **when you were first conceived you were just a single cell, endowed with one master copy of the architect's plans.** *This cell divided into two,*

The DNA Delusion

and each of the two cells received its own copy of the plans. Successive divisions took the number of cells up to 4, 8, 16, 32, and so on into the billions. **At every division the DNA plans were faithfully copied, with scarcely any mistakes...** The coded message of DNA, written in the four-letter nucleotide alphabet, is translated in a simple mechanical way into another alphabet. This is the alphabet of amino acids, which spells out protein molecules. Making proteins may seem a far cry from making a body, but it is the first small step in that direction. Proteins not only constitute much of the physical fabric of the body; they also exert sensitive control over all the chemical processes inside the cell, selectively turning them on and off at precise times and in precise places. Exactly how this eventually leads to the development of a baby is a story which it will take decades, perhaps centuries, for embryologists to work out. **But it is a fact that it does. Genes do indirectly control the manufacture of bodies...**

This is immense news; earthshaking. One of the greatest discoveries in the scientific history of mankind. We finally know what brings about the development of humans from a single fertilized ovum! Modern science and Richard Dawkins say DNA undeniably holds the plans for the human body. Dawkins says those DNA plans have been assembled by natural selection. At the same time, natural selection works on copy errors in DNA coding. Which should make one wonder how the DNA system originated in the first place. It certainly didn't evolve by natural selection, since natural selection needs a genetic system and the DNA it is composed of to work. But let's not quibble about trivial items. Let's celebrate like Richard Dawkins and his friend and supporter Bill Maher are doing in *my* New York Times photo in Fig. 10-4. Maher produced the anti-religion movie *Religulous*, and evolution and the human blueprint support his anti-religion stance. So he was a very happy man. The scientific world must have been stunned. Overwhelmed. This news is as big as the moon landing.

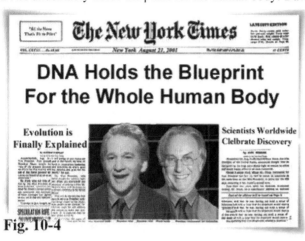

Fig. 10-4

Let's move forward 19 years and see what Dawkins' then has to say about what holds the plans for the human body. The news is even more stunning! The plans *for the entire human body do not* exist in *every tiny cell in our body*! What a stunning and historic reversal! Dawkins' explanation in chapter 8 of his NYT best selling book *The Greatest Show on Earth* (Free Press, Simon and Schuster, 2009), is now the currently accepted scientific model of what directs the development of a fertilized ovum into an embryo and then into a fetus and finally into an infant. Actually I love his new explanation. It makes my book just that much more interesting, because his account is so absurd. It's pure fable disguised as real science. I wonder if I would have fallen for his new story when I was an avid evolution believer. My bet is I would have, since I fell for so many other obvious evo-illusions when I was an enthusiastic evolutionaut. That's what indoctrination does to people, and I am no exception. Only I was cured, while few can be. In *The Greatest Show on Earth* Dawkins discusses preformation as if it's just another silly notion from centuries past. Then he gives the *Modern-True-Scientific-Version*, which is the same: just another silly notion, only a modern version. But in many ways it's worse than the ancient fable. Without any scientific evidence or inquiry, his story completely changes. One would think to make such a change would require earth-shattering scientific testing and investigation. But no testing was done; no research. There could not possibly be any, because there is nothing to test, which makes the modifying of fables easy. In *The Greatest Show on Earth*, Dawkins says:

DNA has a breathtakingly precise way of intersplicing half the paternal information with exactly half the maternal information, but how would it go about intersplicing half a scan of the mother's body with half a scan of the father's body? Let it pass: this is all so far from reality. **DNA, then, is emphatically not a blueprint.**

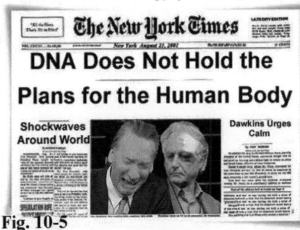
Fig. 10-5

So, only 19 years after declaring that DNA holds the plans for the entire human body, Dawkins declares DNA does *not* hold the plans for the entire human body. This change is earthshattering. It should have been immense news, presented in every major newspaper, journal, and news program on Earth. Time magazine, Scientific American, and NatGeo should have had entire issues dedicated to this unconceivable news like the New York Times did in *my version* in Fig. 10-5. Maher and Dawkins should be devastated. But the news sources of the world were silent. The *new* fact that DNA does not hold the plans for the human body kills off the entire science of evolution. Evolution is based on changes in DNA that form all species, biological systems, and body parts. Isn't it huge news that DNA cannot be *the source*? That the entire science of evolution was dead? One of the main reasons Dawkins cites for his 180^0 reversal is that *intersplicing exactly half the paternal information with exactly half the maternal information* would mean two different plans for the human body would have to be somehow munched together during fertilization. Which means DNA could not hold the plans for a single human body. Of course the information about *intersplicing half the paternal information with exactly half the maternal information* was available to Dawkins when he wrote *The Selfish Gene*. So why didn't he do a bit of critical thinking in 1990 and come up with his same new conclusion? That DNA and our genes do not and could not hold the plans for and direct the formation of the human body from a single cell? Maybe he was fooled by his instructors, just as I was; as was nearly every student of biology on the planet. Dawkins came to his senses, and rejected his explanation for human development he used in *The Selfish Gene*. But again, he blew his chance to be an honest scientist. He substituted his original fable with a new and far more fascinating and preposterous evo-illusion. Dawkins says his new fable is based on another theory called *epigenesis*. The Oxford Dictionary defines epigenesis as:

a theory of the development of an organism by progressive differentiation of an initially undifferentiated whole.

Here is my rule of fables: *as long as the fable has a complex and scientific sounding name, it can be promoted as valid science.* Epigenesis is no exception. Epigenesis involves a ground-up sort of construction, rather than a construction requiring an overall plan or blueprint, and an overseer as in top-down construction. It's as if the bricks themselves build a brick building. Each placed brick "knows" how and where the next brick should be placed. Each brick is responsible for the correct placement and stabilization of the next brick and only the next brick. And that brick "knows" how and where the next brick should be placed; and on and on until an entire building is built. The bricks themselves build the building, one brick at a time, rather than an architect, contractor, and construction workers. There is no overall external guidance. Of course it's impossible to imagine a brick building being built by the bricks themselves because the bricks are inanimate non-living objects. But this notion is transposed to living cells. The "bricks" are the living cells of an embryo. The first "brick" is the zygote, the fertilized egg, which is a single living cell. The zygote next divides into two cells. The information needed by the two daughter cells to form properly and to be properly placed is passed on to the daughter cells by the zygote. Then when the two daughter cells divide into four cells, the two daughter cells become parent cells. They pass on and direct only what is necessary for the four new daughter cells to form correctly, and place themselves just where they are needed for the development of the human infant. Each step continues in like fashion, each cell directing the formation and placement of its two daughter cells, and passing on only the information that the new daughter cells will need until an entire human infant is formed. Where it's impossible to imagine a brick having guidance and direction capabilities, cells are so complex and mysterious, and invisible to 99% of the human population, it's easy to fool the audience into thinking they do have guidance and direction capabilities to form an entire human body from the ground up and from a single cell.

In his book Dawkins makes a huge mistake by trying to equate the bottom-up formation of an embryo with the mythical story of how some medieval cathedrals were built. According to the myth, some cathedrals had no architect or plans to go by. They were simply built by many sub-contractors who plied their trade in the locations of the building where they were needed. Somehow the cathedral was built by these disorganized yet organized subcontractors without oversight, or plans. Somehow miraculously the cathedral came together perfectly, in ground-up fashion. Can you imagine what a structure like this would look like? It would be a useless mess.

Dawkins example is absurd, because his poorly chosen illustration uses subcontractors who organize sections of the building. There is no possible correlation with the formation of the zygote into a full human infant. It would be as if a "subcontractor/organizer" oversaw the formation of the human brain and neurological system, then another oversaw the formation of the musculo-skeletal system, and another oversaw the formation of the cardiac system. That is not what Dawkins himself, or epigenesis espouses. His example has many "overseers" instead of none or one. Bad choice Richard. Further destroying Dawkins' example is the fact that each subcontractor is intelligent. There isn't supposed to be any intelligence in the formation of living nature. So Dawkins picked a horrible example, but as long as it works and fools his enthusiastic readers and the New York Times and NatGeo, no need to worry.

Dawkins' second example of epigenesis and the formation of a human infant is even worse. He uses flocks of starling birds to promote his fable. Starling flocks fly in the most unimaginable and beautiful way. If you have never seen one go to my website:

www.thednadelusion.com (Alternative: www.thednadelusionblog.wordpress.com) Watch video number *9. Starlings*: *Putting On A Show*. You won't believe your eyes. How do these birds form up and fly in the unimaginable and organized flocks that they do? Thousands of these birds fly and swoop and dive in the most incredibly organized fashion. They never collide with each other. There seems to be no leader. They just do what they do. Each bird has to be aware of the other birds above, behind, below, in front, and sideways to themselves. Each bird is programmed to react just perfectly to the other birds nearby. Dawkins notes that starling flocks can be imitated with computer programming. Each starling dot is programmed with "local rules" so that they don't collide with other starling dots. Once properly programmed, the starling dot is copied until there are thousands. The amazing dance of a starling flock can then be imitated on a computer screen. The false assumption here is that because starling flocks can be imitated in a computer, that is backup evidence that shows that an ovum can morph into a full human infant in the same manner that starling flocks fly. Utilizing an intelligent computer technician, plus an intelligently invented and constructed computer and its program to mimic starling flocks only proves that the origination of the programmed dance of the starlings requires intelligence. See? There is no need for **IID**'ers to perform experiments that prove **IID** when they are routinely and naively performed by evo-illusionists. There is no doubt that Dawkins' false starling assumption fools most of his readers and followers, even though he is devastatingly disproving his point. Dawkins is either so intelligent he knowingly is doing so, or he is so naïve he has no idea. Starling flocks have absolutely no connection to and cannot be compared to the embryonic development of a human infant. But Dawkins' multipage discussion of course fools most readers into thinking that they do. Just the presence of the discussion is accepted as evidence when it isn't remotely close to being connected or comparable in any way.

Dawkins states:

The body of a human, an eagle, a mole, a dolphin, a cheetah, a leopard frog, a swallow: these are so beautifully put together, it seems impossible to believe that genes that program their development don't function as a blueprint, a design, a master plan. But no: as with the computer starlings, it is all done by individual cells obeying local rules. The beautifully "designed" body emerges as a consequence of rules being locally obeyed by individual cells, with no reference to anything that could be called an overall global plan. The cells of a developing embryo wheel and dance around each other like starlings in gigantic flocks.

Cells wheel and dance around each other? What a horrible and brainless example of the formation of an infant from a single celled zygote. Dawkins insults human development and human anatomy and our intelligence. He leaves so many holes in his discussion, and so many questions unanswered. It's simply amazing someone supposedly as intelligent as Dawkins could be so inane. For starters, where does the programming for each starling come from? What Dawkins calls "local rules" must be made in his computer. Intelligence and programming are required. What exactly programmed the starling's "local rules"? Natural selection? DNA mutations? The programming and local rules are just there, and don't ask questions.

Dawkins uses something he cannot come close to explaining or understanding, starling programming, as an example for something that he wants to convince his readers that he does understand: what directs human embryonic development. Of course he has absolutely no idea what directs human fetal development, so he turns on the fable machine. Yes, the same one that was turned on in ancient times. Further, the programming of the starlings never changes. Embryonic programming, if epigenesis is correct, must constantly be changed as cells differentiate into the hundreds of different cell types, and form the various organs and biological systems, or all of the different body systems and tissues could never form. Starling flocks are homogeneous. They are all made of the same building blocks: starling birds. Humans, and all multicellular species are composed of hundreds of different cellular building blocks. Starlings do the same dance over and over without a single change in programming. The result is always the same beautiful dance of the starling flock with different shapes occurring every second. Embryonic cells must change their programming over and over in short time spans so that they can make the numerous types of cells and systems needed to make a human. If one were to take thousands of photographs of thousands of starling flocks doing their dances, no two photographs would look alike; no two would have the same shape. In the case of human embryonic development, all organs and biological systems wind up with the same exact shape. If you photographed one thousand livers, or skeletons, or brains from one thousand humans, they would all look exactly the same. There is no variance. The same would be true if you photographed embryonic or fetal organs at the same developmental time. Random shape is a constant and key characteristic of the dance of the starlings. It is not and cannot be with embryonically developing humans or any embryonically developing species.

The biggest problem for Dawkins fable and epigenesis is the fact that, no matter what, a single zygote must carry all of the plans for the entire human body. It then passes on those plans in part as needed to its daughter cells. Each daughter cell then takes only the information it needs to place and produce the next daughter cells, and on and on. No part of the zygote has the capability of holding plans for an entire human; or the plans for any part of a human for that matter. Its genes do hold plans for making proteins, and that's it. Science has discovered and discerned the job of every cell part and organelle, and not one has the capability of carrying and acting on any plans for developing a human. But in any case, the ovum must somehow possess the plans for the entire infant, or there would never be a fully formed infant. Even if the plans and information are divvied up cell by cell according to need, construction, and placement of the next daughter cells as epigenesis theorizes, where are those plans in the dividing cells? What an unbelievable mystery. Dawkins use of starlings as an example of human embryonic development is such an immense failure, it's really useless to spend more time discussing it. There is no correlation.

Dawkins continues his discussion by saying how all life, including life on other planets, most likely *will turn out to have evolved by a process related to Darwinian natural selection of genes.* But he says genes don't hold the plans for the body of any animal. So mutations or changes in genes cannot produce the body parts and biological systems we are all made of and claimed by evo-illusionists. Changes in genes cannot produce anything but bad proteins. Which means mutations cannot not evolve species, body parts, or biological systems of any kind on any world.

Dawkins must have no idea that he singlehandedly destroyed evolution by admitting genes don't hold the blueprints for the human body. Maybe I can have him over for dinner some night, and we can discuss. Wouldn't that be fun?

If the control center of the cell isn't the nucleus, one might challenge me with the obvious question: what about XX and XY sex chromosomes that "determine" the sex of an animal. I would answer that the X and Y chromosomes are *indicators* of gender, and not the *determiners*. Bill Nye, the famous evolution promoting Science Guy had a

Fig. 10-6

children's television show, *Bill Nye the Science Guy*. On his show he did scientific experiments and taught kids about science. In one episode in 1996 he had a little drama to demonstrate how boys and girls are made. In the drama a young girls says:

I'm a girl. Could have just as easily been a boy, though, 'cause the probability of becoming a girl is always 1 in 2. See, inside each of our cells are these things called chromosomes, and they control whether we become a boy or a girl. Your mom has two X chromosomes in all of her cells, and your dad has one X and one Y chromosome in each of his cells. Before you're born, your mom gives you one *of her chromosomes, and your dad gives you one of his. Mom always gives you an X, and if dad gives you an X, too, then you become a girl. But if he gives you his Y, then you become a boy. See, there are only two possibilities: XX, a girl, or XY, a boy. The chance of becoming either a boy or a girl is always 1 in 2, a 50-50 chance either way. It's like flipping a coin: XX you're a girl, XY you're a boy.*

The only problem with Bill Nye's (Fig. 10-6) little drama is there is no possible means XX and XY chromosomes have of controlling the development of male and female sexual apparatus. They can only hold the code for certain types of protein enzymes and hormones needed specifically by males or females. Strips of molecular code used to make proteins, inside of a microscopic cell, do not have the ability to make penises, vaginas, breasts, ovaries, testicles... But, without evidence of any kind, Bill Nye indoctrinated the kids into believing the delusion that they can. Why? He is preparing them for their future indoctrination that they will have when their teachers teach them that random changes in genes made up of DNA, over millions of years, produced all of living nature. In any case, there is no possible way evolution, a goalless, blind, random system could form male sexual apparatus that matches female apparatus. So evolution is eliminated as a source for all of living nature. Of course at this point I have to let you in on a scientific secret told to us by our genetics professor at USC: How do you tell the difference between an X and Y chromosome? You pull down their genes!

There is no doubt characteristics and traits are passed on from generation to generation. And, characteristics and traits do change from generation to generation. There are also very specific patterns to the inheritance of characteristics and traits; again there is no doubt. The entity that controls an organism's cell functions, cell type, and body design, is passed on to its offspring; it's inherited. Even though DNA is not *the* controlling entity, its coding used in the process of making proteins plays a large part in the inheritance picture. DNA is a cog in the wheel of inheritance. But it's not remotely close to being the entire wheel. While DNA and its functions are incredibly impressive and complex, whatever the controlling entity in cells is, it's *infinitely more impressive and complex* than DNA and its functions. Here is a rather obvious list of characteristics that the invisible and as yet undiscovered control center for living organisms must have:

(1) It must be **present in, and exist within the confines of, all living organisms**, be they multicellular, or unicellular.

(2) It must be **individually configured and specifically tailored** to each living organism. It must hold the plans for the *individual anatomical characteristics* of an organism that are not universal to all members of that organism's species; in other words, characteristics that are unique and specific to the single organism. These unique characteristics can then be passed on to the organism's offspring. In humans, the shape and appearance of the face usually matches that of the parents.

(3) It must be **inherited** by offspring from parents.

(4) It must be able to have a conscious **awareness of and control of all activities, movement, and functions of all of the billions of molecules in the internal domain** of all cells of a living organism. The *internal domain* refers to everything within the organism's cell capsule, cell wall, or cell membrane, and is inclusive of the cell wall, capsule or membrane. It must permeate the entire cell and be able to coordinate and control the actions of its molecules. It must be able to control the billions of molecules in a cell during DNA replication as it occurs when a cell is preparing for and going through cell division. It must also be able to control the millions of molecules involved with protein synthesis, and all other cellular biochemical functions and cycles. The cellular control center controls DNA, DNA does not control it.

(5) It must be able to have an **awareness of and be able to control and/or react to all events, activities, functions, and designs of the *external domain*** of all cells of a living multicellular organism. The *external domain* refers to the environment outside of an organism's cell membranes, walls, or capsules. It must permeate all cells of a living multicellular organism, and be able to coordinate the actions of those cells.

(6) It must be able to maintain **communication between the internal domain and external domain** of a cell.

(7) It must **hold the plans for the entire body and all of its individual parts** of a multicellular organism, from the cellular level, to the organ level, to the biological system level. It must hold the plans for the entire cell and all of its individual molecular parts of a unicellular organism, and of each cell of a multicellular organism. It must be able to direct the development of the embryo of a multicellular organism from a single zygote to a mature newborn to a mature adult. It must be able to maintain an accounting of the number of cells needed for each part of a multicellular organism. It must know when to start and stop cell division so that the proper shape and function of all organs and biological systems of multicellular organisms are achieved and maintained during embryonic development and throughout the adult life of the organism.

(8) It must be able to **plan and determine cell type**, and have a mechanism that guides the development of undifferentiated cells into their determined cell types. In other words, it can determine which cells become muscle cells, which are neurons, which are skin cells. In the case of the human body, it must be able to determine all of its more than 400 different cell types.

(9) It must **hold the plans for the *species characteristics*** of an organism that is universal to all individuals in that organism's species. It must be specific and individual to each type of species. In other words, each type of species will have its own control center that is specific to that species. For example, for polar bears, it must hold plans and direct the development of a fertilized polar bear ovum to form first an embryo, then a newborn cub, and then an adult polar bear. It must not just stop at the completion of the full formation of the cub. It continues its controlling when the cub is out of the womb and maturing into an adult bear. Which shows that the control center cannot be in the womb of the mother. It will produce all of the characteristics of polar bears that are specific to polar bear fetal development, and the universal characteristics of adult polar bears. With humans, it must hold human individual characteristics, and produce human intelligence and consciousness that are not present in a newborn.

(10) It must **cease operation instantly upon the death** of the organism.

DNA and its associated molecules are involved with items (1) through (3), and (10). DNA has absolutely no ability to perform any of the other functions. DNA is not capable of controlling any entity inside or outside of the cell. DNA *is controlled* by the control center, as are all other molecules inside of living cells, which includes their cell walls, membranes, and capsules. The fact that all cellular molecules are being controlled can be easily observed and tested. Scientists have identified most molecules and organelles inside of living cells, and not one independent molecule or organelle has the capabilities mandated by (4) through (9) above.

Then, what are the possible remaining candidates in living organisms capable of performing all functions 1 through 10? Astoundingly there is only one: *life* itself. It is known for certain that every item (1) through (10) above does occur. The only entity left that can perform all of the above functions is *life*. There is no other known possibility. Life must be a far more complex entity than any human could imagine. In Webster's 1913 dictionary life was defined as:

The state of being which begins with generation, birth, or germination, and ends with death; also, the time during which this state continues; that state of an animal or plant in which all or any of its organs are capable of performing all or any of their functions; - used of all animal and vegetable organisms.

In other words, Webster's Dictionary of 1913 has no idea what life is. Modern definitions of life haven't improved at all. There is no rational definition of life in existence today. It is known what life does, but what it *is* is a far different matter. Is it some sort of non-molecular invisible gas that permeates every living creature? Is it some kind of invisible spirit? Scientist cannot synthesize life. They can't manipulate it in any way. Life can be terminated, but it cannot be initiated. It must always be inherited from a parent organism. It permeates the entire body of every living organism and every cell. These are fast and unchangeable facts. So what if life was the controlling mechanism that performs all of the tasks above? What if there is infinitely more to life than simply being some sort of mysterious entity that causes organisms to move and react to stimulus? The only known candidate that has the possibility of performing all of the ten functions above is *life*. So I change my mind. I am going to call this **Blume's Theory of Life**, which states:

Life is:

(1) The source of the plans for the body of all living organisms. It directs embryonic development in the case of multicellular organisms. It holds the plans for all individual cells and all of their parts, and directs the formation of daughter cells during cell division. It continues to direct functional and morphological maintenance during the life of each individual living organism.

(2) The source of the control center for all biological processes in the internal cellular domain.

(3) The source of the physical and biological control of the external cellular domain of all cells and all living organism

Blume's Theory on Life is really a basic but immensely broad statement, with complex ramifications. I humbly admit that it fits better than any *known* entity or explanation that biological science has offered so far. I might as well proffer it as my theory, and name it after, ahem, myself. Why shouldn't I take another shot at a Nobel Prize? It would take an immense amount of testing to verify my theory, which undoubtedly will not happen in my lifetime. Modern scientists are too sold on the illusion that DNA and the cell nucleus is the control center and plan holder of all cells. Maybe my theory is completely incorrect. But of all of the possibilities and known entities available as of this writing, I'm betting *Blume's Theory of Life* will someday be validated. The only way I can see it might fail is if someday another massively complex entity that humans have not yet discovered winds up being the control center and holder of the plans for all living organisms. If one is found, it may knock my theory partially or completely out. Which is fine. As long as good science is being accomplished, I will be happy. In any case, the true control center of cells and the holder of the blueprints for the human body need to be located. As long as all of science agrees that it's the cell nucleus and our DNA, this incredible *Puzzle* will never be solved. Locating and understanding the biological control center of living organisms and its modus operandi will possibly be the greatest advance in the history of modern biological science if it is discovered. The ramifications will be enormous. The advances in medicine alone will astound.

Before scientists can claim they have solved the puzzle of inheritance, they first must acknowledge that they do not know what the control center of cells is, and what directs and coordinates groups of cells and embryonic development. That is a first step in correcting a science that has taken an illusory track, instead of a realistic one. Once science has universally accepted that it has no notion what the internal and external control center of cells is, and that it admits it has supported an immense illusion, it can begin work on finding what that control center *really* is. Of course, this kind of admission will be Earth-shaking and embarrassing to the entire scientific community. Science would have to admit that it still has one foot in the Dark Ages, where fables and superstitions ruled. Do I think this will happen in my lifetime? Of course not. Our modern illusions and fables are even more insidious than those of the ancients. Modern biological science has created an immense illusion: the illusion that it knows **how** all biological systems do what they do, and **how** they originated. It fools its audiences because a major part of the illusion is based on incredible scientific knowledge, unlike the superstitions and fables of the past. Our valid knowledge has expanded exponentially thanks to hard working honest scientists. The knowledge we've gleaned from modern scientific research is unimaginable. But humanity's ability and penchant to concoct fables and illusions to explain what is still unknown, which grows with each new discovery, is just as unimaginable.

Chapter 11

Is The Blind Watchmaker Really Blind?

Asking general questions led to limited answers, asking limited questions turned out to provide more and more general answers.-François Jacob

Fig.11-1

The modern "Bible" of evolution is undoubtedly Richard Dawkins famed book *The Blind Watchmaker*. (W. W. Norton & Company, New York, 1987) On the cover (Fig. 11-1) Dawkins says his book will let the reader know:

Why the evidence of evolution reveals a universe without design.

A few of my favorite Richard Dawkins quotes are inside this treatise on a **D**umb **L**uck universe. Let's see if he really deep down inside believes the universe is without design. On page 301 Dawkins discusses the eye:

Think of all the intricately cooperating working parts (of the eye): The lens with its clear transparency, its colour correction and its correction for spherical distortion, the muscles that can instantly focus the lens on any target from a few inches to infinity; the iris diaphragm of "stopping down" mechanism, which fine-tunes the aperture of the eye continuously, like a camera with a built-in light meter, and fast special-purpose computer; the retina with its 125 million colour-coding photocells; the fine network of blood vessels that fuels every part of the machine; the even finer network of nerves- the equivalent of connecting wires and electronic chips. Hold all this fine-chiseled complexity in your mind and ask yourself whether it could have been put together by the principle of use and disuse. The answer, it seems to me, is an obvious "no".

I must include, he astonishingly reverses himself and goes on to refute his own "no", and comes up with a "yes". He thinks a hyper-complex digital camera like the eye that he so eloquently describes can form from thousands of completely unlikely and absurdly complex **D**umb **L**uck events that are chosen by natural selection. He also thinks that just the correct number of eyes, two, necessary for depth perception, could and did form from these random **D**umb **L**uck events. In his discussion he skips the need for an optic nerve, and visual cortex in the brain that translates chemical signals into perceived images; both completely necessary for vision to occur. Does he

The DNA Delusion

really think vision came into existence from **D**umb **L**uck happenstance? From this statement I would say deep down he has major doubts.

On page 24 Dawkins discusses the mind-bogglingly engineered bat sonar:

Bats are like miniature spy planes, bristling with sophisticated instrumentation... The mounting and hinging of these three bones (sonar bones) *is exactly as a hi-fi engineer might have designed it to serve a necessary "impedance-matching" function... Echo-sounding by bats is just one of the thousands of examples that I could have chosen to make the point about good design. Animals give the appearance of having been designed by a theoretically sophisticated and practically ingenious physicist or engineer...*

The word *design* just reeks of intelligence. Bat sonar looks designed, it works marvelously which takes intelligence, but it's not intelligently designed? Dawkins goes on to marvel at how bats must, in hyper-rapid cycles, turn on and off their sonar so echoed signals can be picked up by their sonar receiving mechanisms. He says it's much like a World War I fighter plane that had the ability to shoot through their propellers. The timing is critical and designed into bat sonar. So Dawkins himself proves design in his book; but then he goes on to reiterate there is no design. He must be a very confused person.

Here is one final Dawkins quote. He hates this one the most. It has come back to bite him more than any other. But once the cat is out of the bag…

The Cambrian strata of rocks, vintage about 600 million years, are the oldest ones in which we find most of the major invertebrate groups. And we find many of them already in an advanced state of evolution, the very first time they appear. It is as though they were just planted there, without any evolutionary history.

Dawkins has made more excuses for this quote than a baseball batter going through a yearlong slump. But he said it, he thought it. And once said, once printed in millions of copies of his book, it might as well be engraved in stone.

Dawkins gets into a major discussion on the evolution of proteins. He examples the hemoglobin molecule, because it is an average sized protein. The hemoglobin molecule is made up of four strands of amino acid molecules. Two strands have 141 amino acid molecules each, and the other two have 146. Kind of like the pearl necklace I exampled earlier. Only hemoglobin has four separate attached strands of amino acid molecules crisscrossing each other. The total number of amino acid molecules in hemoglobin is thusly 574. Dawkins discusses the odds of a hemoglobin molecule forming randomly in some location somewhere. He doesn't give the location of the formation, so it must be assumed the assembly takes place inside of some random **D**umb **L**uck cell. In his discussion, he casually dismisses three of the strands by saying, "Let's just think about one of these four chains". Actually, when I first read his book, like probably all readers, because this statement comes before his discussion, I didn't know or realize what he was getting at. I completely dismissed and missed the huge meaning of his sentence, "Let's just think about one of these four chains". Here is evo-illusion at it's best. His ability to sneak in "Let's just think about one of these four chains." and dismiss the other three strands without raising great scientific eyebrows is astounding, and really a top-notch evo-illusion. So kudos to you Richard. Eliminating three of the four strands greatly increases his upcoming phony odds calculations in evolution's favor. Dawkins cheating actually matters not. Let's let him do his evo-illusion, his cheating, and work with *his* numbers. Figuring the odds of only one strand forming whilst ignoring the other three will do the job for me just fine Richard. Using Dawkins' own evo-illusion will kill the notion of the random assembly of hemoglobin; and the evo-illusion that goes along with it. So as Dawkins says, let's just dismiss the other three strands.

Dawkins calculates that the odds of one single hemoglobin molecule *strand* forming randomly when the proper building blocks are intelligently and mechanically isolated and in close proximity in some sort of beaker or closed system container is 1 in 10^{190}. In case you forgot your math, 10^{190} is a one with 190 zeroes after it. To give you an idea of the absurdity of this number, one trillion has 12 zeroes. 10^{190} has an unusual name: *ten duosexagintillion*. It's amazing that anyone would assign a name to such a useless number. When hemoglobin first formed from **Dumb Luck** as Dawkins says it did, each location on Dawkins single strand had a 1:20 chance of locking in the correct amino acid, as there are twenty different amino acid types used by vertebrates. Actually the number of amino acids in living organisms is twenty-two, which would make his hemoglobin odds about $1:10^{196}$. This would make everything much worse for Dawkins' evo-illusion, so we'll stick with twenty. One would figure the odds of 146 amino acid molecules mounting randomly but correctly by multiplying 1:20 x 1:20 x 1:20... 146 times. The result is $1:10^{190}$. There are approximately 10^{80} or *one hundred quinvigintillion* atoms in the universe, to give you an idea of the chances of any protein molecule forming from randomness. 10^{80} is only $1/10^{110}$ or one *one hundred quintrigintillionth* of 10^{190}! When working with numbers such as these, discussions are useless. You are essentially working with infinities. The odds $1:10^{190}$ are only valid if there truly is an affinity for amino acid molecules to link, which there isn't, and if the exact and proper ingredients are placed in close proximity, which they could not be, without an intelligent lab tech doing the placing. Selection of ingredients and complete isolation of those ingredients requires intelligence. Without selection and isolation, the odds would not be remotely close to $1:10^{190}$. They would be much much worse. Isaac Asimov was so stunned by the odds of $1:10^{190}$, he gave them the name *The Hemoglobin Number*. He should have titled it *One Fourth of the Hemoglobin Number,* as Dawkins uses only one of four strands. There are millions of proteins utilized by living organisms. The random formation of each protein chain would have similarly absurd odds. For the fun of it, and for Richard, I have calculated the odds of all four strands randomly forming to make up one single molecule of hemoglobin. The answer, and the *True Hemoglobin Number* is $1:6.2 \times 10^{746}$ or:

one in six hundred and twenty duocenseptenquadragintillion.

Since there are 10^{80} atoms in the entire universe, I think you can see where we are. 1.6×10^{746} is more atoms than there would be in all of 10^{666} (*one duocenunvigintillion*) universes. Richard got away with a fantastic evo-illusion here. Proteins are made of chains, like snap-together pearls. The notion that any amino acid molecule would have any chance of finding its way into position on a partially assembled "pearl necklace" of amino acid molecules is absurd. Just think, it would have to break the bonds of two previously bonded amino acid molecules, push them apart and snuggle itself in, then make two new bonds. Absurd to even think about. But I will discuss it like Dawkins does, as if it could happen.

In his book *The Blind Watchmaker*, in order to build on his illusion and to solve this seemingly unsolvable problem for evolution, Dawkins asks that his readers do a mind experiment which hopefully will make the horrendous odds of the hemoglobin number seem like it can be overcome. Can you imagine trying to explain away $1:10^{190}$? This could only be attempted by an evo-illusionist. Dawkins asks his readers to visualize monkeys typing randomly on typewriters until one comes up with this sentence: METHINKS IT IS LIKE A WEASEL. There are 28 letters and spaces in this Shakespearean phrase. There are 26 letters in the alphabet, and with a space the total possible digits is 27. So each of the 28 locations has a 1 in 27 chance of having the correct letter or space correctly typed. To calculate the odds of METHINKS IT IS LIKE A WEASEL being randomly typed by monkeys banging on typewriters, one would have to multiply 1/27 twenty eight times. The odds come out to $1:10^{40}$ or a one in a one with forty zeroes after it. Or *ten duodecillion*.

Instead of writing the odds $1:10^{40}$ out numerically like he wrote the Hemoglobin Number, $1:10^{190}$, Dawkins changed his *modus operandi* and wrote it out verbally: "1 in 10,000 million million million million million million". Writing it out verbally instead of numerically impresses his readers, of course, into thinking he picked a task with absurdly low odds like the $1:10^{190}$ for the chance assembly of one strand of hemoglobin. 1 in 10,000 million million million million million million sure seems as absurd as $1:10^{190}$. Why didn't he write it out numerically so the reader can honestly compare the two? Actually most readers wouldn't know the difference, so he sure would have been pretty safe writing $1:10^{40}$. The odds for the task he picked, $1:10^{40}$, are near infinitely low and *against*, but still immensely great in comparison to the odds for the random self-assembly of one strand of hemoglobin. His 1 in 10,000 million million million million million million isn't remotely close to $1:10^{190}$. Does he realize this? Did he take math in school? My bet is he sure as hell did realize. And he certainly is math trained. He actually states, "It is the same kind of calculations as we did for hemoglobin, and it produces a similarly large result." (p. 47) "Similarly large results" is flat out wrong. Dawkins should know that using powers of 10 is not at all like counting apples and oranges. For instance, subtracting 10^3 from 10^9 yields 9.99999×10^8. Just ever so slightly less than 10^9. It's interesting to note how numbers can fool people, and Dawkins certainly uses his numbers to fool. To determine the difference between 10^{190} and 10^{40} by subtracting, the result is astonishing. It's still about 10^{190}! Here is the exact result to 149 decimal places:

9.99999999999999999999999999999999999999999999999999 99999999999999999999999999999999999999999999999999999999 9999999999999999999999999999999999999999999999999 x 10^{189}

You see, you're still left with almost 10^{190}. But that's typical evo-illusion; the evo-illusionists fooling their audiences and whoever else will listen. If Dawkins used a phrase with 95 or 96 letters and spaces, such as the last phrase in the Gettysburg Address,"...THAT THE GOVERNMENT OF THE PEOPLE, BY THE PEOPLE, AND FOR THE PEOPLE, SHALL NOT PERISH FROM THE EARTH", he could have had a more appropriate comparison. The odds of monkeys randomly typing Lincoln's phrase are about $1:10^{190}$, the same as the hemoglobin number.

To immensely improve the odds for the monkeys randomly typing Dawkins' chosen phrase, Dawkins uses what he calls *cumulative selection*. He made a computer algorithm that randomly placed letters in a previously constructed template composed of his target WEASEL phrase. Of course no template exists in nature. But one exists in Dawkins' evo-illusion. In his algorithm, each time a letter landed in the correct location needed to form the WEASEL phrase, it is "cumulatively selected". It remains in place, instead of completely reloading the entire sentence each time the "monkeys" in his algorithm typed a mostly incorrect 27 letters and spaces. In Dawkins' world, the monkeys typing correlate with hemoglobin trying to randomly invent itself. With Dawkins' slick cumulative selection illusion, each jumbled molecule with some correct letter placements was supposedly saved in the digital organism that was in the process of "evolving" his phrase. That organism then has offspring, which allowed for another round of cumulative selection. Each new generation did the same, saving the good placements and only allowing the reloading of the incorrect ones. Generation by generation METHINKS IT IS LIKE A WEASEL was eventually and easily formed. By keeping the correct letters accidentally placed in their correct position, then adding to those, Dawkins says the odds of a monkey typing METHINKS IT IS LIKE A WEASEL and hemoglobin evolving in organisms get much more favorable. Dawkins template means he is playing Wheel of Fortune with hemoglobin, a game that takes intelligence to set up.

As I said earlier, proteins are chains, much like our plastic snap-together pearl necklace in my musical chairs game. Spaces left in Dawkins' chain of letters are just fine in a computer algorithm. The letters can snuggle themselves into place. No snapping together needed. Incorrectly placed letters were kicked out of his template, and the spaces were saved for the future correct placement of letters. In real life molecular chain assemblages, there wouldn't be a template to hold pieces together unless one was intelligently constructed. A few amino acids segments that might have, by Dumb Luck, correctly linked together would just float away. There would be no template holding the amino acid molecules in place, as was the case in Dawkins fake algorithm. His mind experiment is a massive fraud. It was obviously and transparently put together to fool the masses; which he did successfully. Of course, again, the template used by Dawkins' imaginary monkeys supposedly correlated to the correct formation of a protein molecule, which did not previously exist and had not yet been invented when the first hemoglobin molecule supposedly first evolved. This requires intelligence and pre-planning. Poor Richard. He keeps introducing intelligence when he says there was none involved with the origin of living organisms.

Dawkins should call his process *cumulative cheating*. Cumulative selection *is* cheating, but that matters not. This is Dawkins' evo-illusion. Illusionists cheat, and that is how they fool people into believing they are seeing what they are not actually seeing. If Dawkins is a golfer, I wonder if he uses cumulative selection on the golf course. He could quickly be a scratch golfer no matter how bad he may be. It would work even better with bowling. Under cumulative cheating, in each frame, Dawkins would get as many balls as he needs to bowl a strike. He could keep the frames that have strikes, and redo the frames that aren't strikes until they are. That way he could roll perfect 300 games by the dozen! And become the best bowler in the world! When Dawkins ran his algorithm on a computer, with the WEASEL phrase typed in as his target, he came up with METHINKS IT IS LIKE A WEASEL in 64 generations and about a half an hour. On another try it was only 40 generations. Dawkins didn't emphasize and clearly explain in his book that his programming METHINKS IT IS LIKE A WEASEL into his algorithm as a target for the cumulatively selected letters to aim for has absolutely nothing to do with the odds of hemoglobin forming from evolution and **Dumb Luck** when no template or goal exists. He explained it as if his algorithm was a good rationalization for the hemoglobin number, when he had to consciously know it wasn't. **Dumb Luck** and evolution have no targets. What Dawkins actually did is prove **I**ngenious **I**nvention and **D**esign (**IID**). His own intelligence and planning were needed to get his computer to randomly come up with METHINKS IT IS LIKE A WEASEL. I wonder why Dawkins didn't just fill a 55-gallon drum with all twenty amino acids and see if any hemoglobin molecules form out of **Dumb Luck**. Wouldn't that have been a better test than his bait and switch "monkeys typing" algorithm? Actually Dawkins and I know the reason: he wouldn't get a single protein molecule of any kind.

If his computer program corresponded to a hemoglobin molecule randomly forming in a living organism, the organism in which hemoglobin was evolving would have had a fully formed hemoglobin molecule template as a target model for Dawkins' cumulative selections to aim for. That just simply isn't possible, unless some intelligent inventor seeded hemoglobin or the plans for hemoglobin into the organism in some fashion, then instructed the cell to copy the implanted hemoglobin molecule, or to make a molecule according to the plans. The cell would have had to "know" that hemoglobin was its target.

Another big question here is why Dawkins didn't write his algorithm using the *actual* DNA code for hemoglobin in the first place. Each amino acid could have been assigned its own actual codon. Each codon for each amino acid is made up of three letters, not just one. Using actual coding would have been far more accurate, and it would have worked much better than the silly WEASEL phrase he used. The actual DNA coding would have required Dawkins to multiply the number of amino acid molecules in hemoglobin (574) times the number of letters in a codon (3) or 574 x 3=1722 letters. A short segment of the 1722 letter coding would look something like this:

...AGTTCTCAAGCTTACAGACTAGTTCTCAAGCTTACAGACT...

So to calculate the odds of a single hemoglobin molecule randomly made by a DNA coding system would mean multiplying one over four, the number of different coding letters (¼) 1722 times. The result would be $1:5.5 \times 10^{1036}$. So now you know why Dawkins didn't use DNA coding to calculate his odds instead of monkeys typing the METHINKS IT IS LIKE A WEASEL phrase. The answer is rather obvious. The WEASEL phrase allows him cheat. He baits and switches to distract not only from the real odds of the random assembly of hemoglobin, but from the fact that each amino acid molecule is represented by its three letter codon, not just one single letter as in the WEASEL phrase. Using the actual code would certainly make more sense, but it would be a reminder to the reader that the DNA code could not have been invented and assembled by a process as simplistic as evolution. Distraction is a typical *modus operandi* for all illusionists; evo-illusionists are no exception. Dawkins was cheating on top of cheating. He cheated by eliminating three of the four strands of hemoglobin. He cheated by using his far simpler WEASEL phrase. He cheated by inserting a template when he knows there was none. He cheated by inserting cumulative selection. He cheated by ignoring so many other variables that needed to be addressed. Did Dawkins know what he was doing when he cheated? Of course he did. Has he been called out on his cheating by good objective scientists? Of course not. At least not that I have ever observed.

So Dawkins is the perfect example of an evo-illusionist. Here he performed an evo-illusion for his readers, and most of them certainly fall for it. I did the first time I read his book. I had to go over it several times to unearth the illusion. If he used a task with the actual odds of $1:10^{190}$ like monkeys typing the last phrase in the Gettysburg Address, I kind of doubt a half hour lunch break and 64 generations would be sufficient to randomly form his goal even with his cheating technique of using a template and cumulative selection. At the rate that resulted from Dawkins' lunch break and algorithm, it would take 5×10^{149} (*five octaquadragintillion*) one-hour lunch breaks, or 5.8×10^{145} (*fifty-eight steptenquadragintillion*) years for monkeys to randomly type the last phrase in the Gettysburg Address; even with cumulative cheating. Just think of how many hamburgers Dawkins could eat in that span of time.

Of course many questions arise out of Dawkins mind experiment. What is so much fun about Dawkins' experiment is that he proves **IID**, and disproves evolution. As is always the case with evo-illusionists, he so innocently proves what he is trying to disprove. A biochemical soup either inside or outside of a cell capsule half a billion years ago would have had no hemoglobin target for amino acid molecules that may have been hanging or floating around. Freely floating amino acids molecules could not have had foreknowledge of what the purpose and design a particular target might be.

In mathematics the number $1:10^{40}$ (*ten duodecillion*) is considered the demarcation line for a complete and infinite impossibility. Mathematicians, as a rule of thumb, won't deal with figures beyond 10^{40}, which makes this discussion and all of Dawkins effort moot; so here we are, arguing in the realm of the absurd. Actually Dawkins cheating makes for such a fascinating puzzle that is so much fun to analyze, so I really don't mind a bit.

There are many other factors that make the odds of $1:10^{190}$ far worse for evolution, if that's possible. Note that Dawkins didn't mention any of these in his book. I find it astounding that he didn't. Did he forget? Of course not. This is just another part of his evo-illusion. It would be like an illusionist doing the linking rings trick and showing his audience the opening in his ring. Here are only some of the other factors that Dawkins didn't bring to the table.

(1) Timing: Hemoglobin had to form in some organism that had already evolved the equipment that made the organism capable of utilizing it. In other words, the organism had to have already evolved red blood cells that carry hemoglobin. So did the organism that evolved one single hemoglobin molecule also already have blood cells that carried that one hemoglobin molecule to some cell somewhere? Did the organism also already have blood vessels? A heart of some kind to pump that one molecule to the location on the organism that needed oxygen? Because if it didn't, the hemoglobin molecule would have been worthless and useless. Timing was everything. Actually, a single hemoglobin molecule, assembled against the odds of 1 in 10^{190} would have been worthless anyway. Billions of molecules would have been needed even in a flea to do any good at all. A single hemoglobin molecule would have been a miracle of **Dumb** Luck that would have been 100% useless.

(2) **Location** in the first organisms that evolved hemoglobin: Did that one incredibly valuable first single hemoglobin molecule form inside of a red blood cell? If it came together, say, in the feces of the organism, or outside of any entity that could transport it to an organ that oxygenated it and could transport it to where cells needed it, the molecule would have just disappeared or wound up floating off in some body of water somewhere. If the molecule deteriorated before it could be used by the host, the clock would have had to start all over on another hemoglobin molecule. Luckily there was lots of time, and according to Dawkins, anything can happen with lots of time. Right?

(3) **Genome modification:** Did the organism that won the Lotto of the Universe and of Infinity by randomly forming a hemoglobin molecule quickly code its DNA for that one molecule so more molecules could be manufactured? Or did it have to wait until another $1:10^{190}$ event occurred and a second molecule showed up? Then a third...? Then...? Did *supernatural selection* act quickly enough to save that single molecule? It must have, because we all have lots of them. Or was the DNA itself in the first organism that formed hemoglobin trying to make some kind of protein that would carry oxygen? Did DNA have foreknowledge of what was needed? This is really the grand-daddy of all problems for Dawkins and his hemoglobin fable. He doesn't dedicate a single line in his book to say how the **Dumb** Luck hemoglobin molecule became coded in the DNA of the organism that first formed it. I wonder why.

(4) **Many different proteins:** Was one hemoglobin molecule enough to save some cells with its oxygen, and allow supernatural selection to select the super-lucky organism with that one molecule so that future organisms would be lucky also and evolve hemoglobin? Because Dawkins and other evo-illusionists talk as if that's all that was needed to set off an incredible biological series of miracles. Dawkins talks of a single molecule of hemoglobin; his algorithm involves a single molecule. He sweeps under the rug that there are millions of kinds of proteins that need to be invented by evolution, and that the odds of each one chance assembling is just as absurd as the odds for hemoglobin. The story is the same for each and every protein. Proteins average about 500 amino acids each. Dawkins needs trillions of trillions of monkeys typing for trillions and trillions of years. Actually, that wouldn't even do the job of manufacturing one single hemoglobin molecule.

(5) **Migration to other species:** How did that single first hemoglobin molecule spread from the first species that evolved it to other species? Since species cannot

interbreed, it would seem that hemoglobin molecules would have had to form independently in many other species. In fact all animal species independently had to form their own hemoglobin molecules, or find some really unusual and really ingenious technique to get the stuff from species that have it; a technique that we humans sure aren't privy to. Hey, maybe that one molecule evolved in the common ancestor of all animal species. And all other protein molecules did as well. Can you imagine the odds?

(6) Odds of a second molecule, and a third... : Did Richard Dawkins take into consideration the odds of a second hemoglobin molecule forming after the first one formed? In figuring out odds, let's say there are two unlikely events, each with a 1:10 chance of occurring. You can calculate the chance of both occurring by multiplying 1:10 times 1:10. The odds of both events occurring are then 1:100. Transposing that to the hemoglobin molecule, if the chance of one molecule randomly forming is $1:10^{190}$, the chance of a second hemoglobin molecule forming is $1:10^{380}$. The chance of a third molecule forming is $1:10^{570}$. It really just gets absurd with only three molecules, even using Dawkins' fraudulent odds. Remember, there are 10^{80} atoms in the entire universe. Even with Dawkins selective cheating, his odds would have been reduced to beyond impossible quickly. If his odds were $1:10^3$ for evolving one molecule, evolving 100 molecules by chance assembly would have dropped to $1:10^{300}$. I really wonder why Dawkins didn't consider multiple molecules himself. Or maybe he did. All of the poppycock about monkeys typing is a total waste of time. The requirement for a second molecule and the third kill the monkeys typing argument.

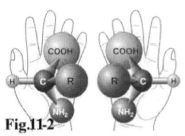
Fig.11-2

(7) Chirality: Richard doesn't take into consideration the fact that every amino acid molecule that hooks up to other amino acid molecules is a left-handed molecule Amino acids have both right and left-handed mirror image versions, kind of like your hands have right and left handed versions. (Fig. 11-2) To give you an idea of what I am discussing, think of trying to make a necklace out of all white pearls. You have a vat of mixed up black pearls and white pearls, 50% of each type, as your source of pearls. Your challenge is to blindfold yourself and randomly pick out and assemble a string of pearls that are all white. The odds of randomly selecting 146 white pearls to make the necklace would be calculated by figuring 2 to the 146th power. The resulting odds of forming one single chiral protein strand by randomly selecting from a vat of right and left handed amino acid molecules are $1:9 \times 10^{43}$ (*one in nine tredecillion*). Yes the odds would be much worse than the $1:10^{40}$ odds of Dawkins monkeys randomly typing his WEASEL phrase. I wonder if Dawkins knows what chirality is. As important as it is, he doesn't mention it in *The Blind Watchmaker*. I wonder why. Charlatans never mention anything that might ruin their schtick.[1,2]

(8) Bonds: Amino acid molecules can bond together like a necklace using two different kinds of bonds. The most frequent type of bond is a peptide bond. The less frequent bond is a non-peptide hydrogen bond. When proteins are synthesized in our cells, their amino acid molecules must be connected by peptide bonds 100% of the time. If they connect with hydrogen bonds they would be useless. How would the "randomness" of evolution produce 100% of anything?[3]

(9) Geographic Location: If a second hemoglobin molecule formed on its own, did it form in the same geographic location as the first? What if a first molecule formed by the North Pole, whilst the second molecule formed near the equator! Actually if a molecule of hemoglobin formed only one foot away from another, it would be similar to two people being 5800 miles apart! Would the two molecules that formed out of **Dumb Luck** locate each other so they could work as a team of two, and be recognized by and carry oxygen for the organism that originated them?

(10) Assumption that amino acids would join: For the hemoglobin number to even have validity, it must be assumed that free amino acid molecules were frantically floating around, trying to join with other amino acid molecules to make something; hopefully hemoglobin. Why wouldn't they just remain as freely floating molecules in whatever environment they were in? Why would they be frantically joining together at all? In Dawkins mind experiment, the monkeys have a task: that of typing. In reality the notion of amino acid molecules coming together and forming proteins would mimic the scene of putting one thousand monkeys in a room with one thousand typewriters and hoping they will all sit down and type. The monkeys would do anything but type. Reality is it would be complete monkey chaos. The monkeys would probably take great big craps on the typewriters, or throw them on the floor. In a world without intelligence, organization, or motivation, it's highly doubtful that amino acid molecules would be looking to do anything but be amino acid molecules; which makes the hemoglobin number $10^{infinity}$. Amino acid molecules don't have a great affinity to link up. [4]

(11) Invention: Dawkins completely ignores his biggest problem of all: that of the invention of hemoglobin. Just think of how incredible this invention is: a protein that carries oxygen and yields it easily to oxygen-starved cells. On the Early Earth of 500 million or so years ago, was hemoglobin new? Useful? Not obvious? Of course it was. *New, useful, and not obvious* are the criteria for determining whether any entity is patentable by the United States Patent office. Under these criteria, hemoglobin would have been accepted as an invention without the slightest protest. **IID** would have been listed as the inventor. Dawkins certainly proved that random mutations and natural selection had absolutely zero chance of being the inventor or the designer. The notion that the Early Earth, with absolutely zero IQ, could come up with the idea of hemoglobin is just so absurd. Actually, that's the case for the first version of each and every protein in each and every cell; and the first version of each and every biological system. Each and every eye, each and every pump (heart), each and every brain, each and every ball and socket joint, each and every auditory system, muscle, filter (kidneys, liver), bird nest... The first version of each and every utilitarian entity in living nature is an invention, pure and simple. Invention is the real *coup de gras* for evolution. Inventions require intelligence, and cannot be formed without it.

(12) Not just a chain: Hemoglobin isn't just like a straight pearl necklace. It has a very complex structure that wipes out any notion Dawkins might have for the chance assembly of a hemoglobin molecule. The odds of self-assembly are far less than $1:10^{190}$. To give you a vague idea of how complex a hemoglobin molecule really is, here is a small segment of a description of its design. Feel free to glance at this paragraph, or skip it entirely. All you need to know is that hemoglobin isn't just a chain as Dawkins touts and needs for his monkeys algorithm.

Hemoglobin has a quaternary structure characteristic of many multi-subunit globular proteins. Most of the amino acids in hemoglobin form alpha helices, connected by short non-helical segments. Hydrogen bonds stabilize the helical sections inside this protein, causing attractions within the molecule, folding each polypeptide chain into a specific shape. Hemoglobin's quaternary structure comes from its four subunits in roughly a tetrahedral arrangement. In most vertebrates, the hemoglobin molecule is an assembly of four globular protein subunits. Each subunit is composed of a protein chain tightly associated with a non-protein heme group. Each protein chain arranges into a set of alpha-helix structural segments connected together in a globin fold arrangement, so called because this arrangement is the same folding motif used in other heme/globin proteins such as myoglobin. This folding pattern contains a pocket that strongly binds the heme group.[5]

(13) A reservoir composed of only hemoglobin amino acids: For the evolution of hemoglobin to have taken place, there needed to be some kind of reservoir that contained only and all of the amino acids that occur in hemoglobin so that at least one hemoglobin molecule could be randomly assembled. This reservoir had to be genetically passed on from generation to generation as the hemoglobin molecule was building itself. What if only five of the needed amino acids were present in the reservoir? Or what if many other biochemicals were present in the reservoir that weren't needed in the formation of hemoglobin molecules? Of course they all had to be left-handed molecules.

(14) To put the final nail in the coffin: There are about 4.23×10^{17} seconds in 13.7 billion years, the time since the beginning of the universe. There are about 10^{80} atoms in the universe. If each and every atom went through some sort of chemical or nuclear event every second since the beginning of the universe, the total number of events would be 4.23×10^{97} events. This doesn't even represent impurities compared to the chance formation of Dawkins' one strand of hemoglobin: $1:10^{190}$. Subtracting the total number of atomic events that could occur if something happened to each and every atom each and every second since the beginning of the universe from the probability of hemoglobin forming from some sort of biochemical soup, $1:10^{190}$, yields almost $1:10^{190}$! Here is the exact number to 174 decimal places:

9.999999999999999999999999999999999999999999999999999999999
9999999999999999999999999999999999999999999999999999999999
9999999999999999999999999999999999999999999999999999999999
999999557 x 10^{189}

This discussion is so ridiculous. Why is it even necessary? The odds of the chance formation of proteins alone kill the notion that evolution is responsible for all of living nature. With only this single argument, evolution and Dawkins should curl their tails under their legs and go off and hide. But they won't, so evo-illusionists like Dawkins and absurd discussions like his will continue on until Dawkins and his fellow evo-illusionists get honest and admit the illusion.

Chapter 12

The Abio-alchemist Stirred the Pot and, and...

The more I examine the universe and study the details of its architecture, the more evidence I find that the universe in some sense must have known we were coming.- Freeman Dyson

So how do modern scientists say the first DNA molecules formed? Of course they had to appear before living cells appeared. There is no possible way cells and all of their billions of parts including DNA could have appeared together through **DL**. Do modern scientists have a rational theory about the origin of DNA, life, and all biochemical systems? When listening to abio-alchemists talk about the birth of life on our beautiful blue planet, they seem so scientific; so confident; so logical. Until you use your head and think out what they're actually telling you. Is their scenario for the beginning of life possible? With a bit of high school logic and mind experimenting, it's easy to see that the most commonly accepted scientific scenarios fail very quickly. In fact, there is no plausible scientific scenario that any man who has ever lived on the face of the Earth can or has come up with that explains the first living cells. Take a moment and think of how incredible that is. No person has *ever* come up with any kind of plausible scenario. That should render a notion about how close we humans are to figuring out our origins and the origin of life itself; how close we are to solving *The Puzzle*. Evolution fails scientifically right out of the starting blocks. And it doesn't get back on its feet anytime during the race. Evolution is dead in the seas of the Early Earth.

Modern science in the form of evolution is absolutely certain it knows how all species and bio-logical-systems formed. Kind of in the same way it was sure it knew how all of life formed through *spontaneous generation,* an early and disproved theory that proposed if you put the right ingredients together, life will spontaneously form on its own from sterile muck. The crash of spontaneous generation was a rug pulled out from under evolution. Evolution will have its own rug pulled out from under in the same fashion. Spontaneous generation, gave way to other equally absurd theories and notions. Reality is we just don't know how the first living cells formed, and there is no plausible scientific theory.

Since there has been such frustration by scientists trying to synthesize life, as I said, evo-illusionists have distanced themselves from the subject. Modern scientific fable-makers have simply moved spontaneous generation, which cannot be accomplished by intelligent lab techs and all of their modern scientific equipment, to billions of years ago when there were no lab techs. According to the fable, **Dumb Luck** put the chemicals together that made living organisms. The imaginary results couldn't be observed or tested. Billions of years replaced lab techs as the maker of DNA and living cells. Evo-illusionists have moved spontaneous generation to billions of years ago. Since it isn't performed by lab techs, the name has been changed to protect the fable. It is now called *abiogenesis*. I call it *evo-abiogenesis* because it is inexorably tied to evolution. It is the beginning, the foundation of evolution. Scientists say it is directed by random lucky steps, what I call **Dumb Luck (DL)** and time. Evo-illusionists tout that abiogenesis has nothing to do with evolution, the gradual change of individual cells into multicellular organisms and ultimately to humans. They completely exfoliate abiogenesis from evolution. Why? Because they obviously have no explanation for the advent of life and living cells on Earth. In reality, the formation of life's first cells is the most important part of evolution. Evo-illusionists putting in so much effort to try to separate the beginning of life from evolution is just another part of evo-illusion.

Evolution scientists jump through hoops trying to convince people that an immense number of infinitely and astronomically unlikely steps took place in going from early simple chemistry to *proto-cells* or pre-biotic cells. Science's answer for the formation of life lies somewhere in between "we have absolutely no idea" and "here is a completely implausible notion (gulp!). We hope you believe it." Spontaneous generation cannot happen today when lab techs put all of the perfect ingredients together in an environmentally perfect lab, but they want you to believe it did happen randomly and accidentally billions of years ago in the hot steamy cauldron of the Early Earth. They get you to believe, and disguise the illusion, by changing the name from *spontaneous generation* to *abiogenesis*.

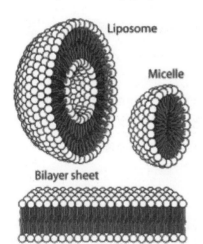

Fig. 12-1

Cells could not exist without the capsule that surrounds them, just like a vitamin with gooey ingredients cannot exist without its vitamin capsule. The earliest cell capsules were supposedly *micelles*, tiny bubbles that are composed of *lipid* (fat) molecules. One end of the molecules that make up micelles is *hydrophilic* (attracted to water), the other *hydrophobic* (repellent to water.) Due to this fact, lipid molecules can line themselves up in water to form perfect little spheres with the hydrophilic ends outside, and the hydrophobic ends inside. (Fig. 12-1) Cell capsules had to be the first structures, the first step, in the foundation of the first proto-cells, no matter what they were made of. Otherwise there could be no cell organization or formation. All parts would float off into the brine of the Early Earth. Cell capsules had to be the first step; much like laying a foundation is the first step in the building of a house. [1,2]

According to evo-abiogenesis, these first micelles that lead to *proto-cells*, the first living cells, would have had to form in the late *Hadean* eon or early *Achaean* eon, around 3.8 billion years ago. The *Hadean* was so named because the Earth would have been a very hot and toxic place; a hellish environment for life to form. Of course the term *Hadean* takes its name from hell or Hades. The Earth was so hot there were no oceans, streams, or ponds. Earth was a dry hot sterile planet that resembled the surface of Venus or the sunward side of Mercury.[3]

If you were able to time-travel back to visit the Earth during the early Achaean (3.8 to 2.5 billion years ago) you would certainly not recognize it as the same planet we humans now inhabit. Worse yet, you would die almost instantly without major protection. The atmosphere was very different from what we now breathe. At that time, it was probably an atmosphere of methane, ammonia, water vapor, and other gases, which would be toxic to all life on our planet today. Yes, incredibly, the first cells formed in an environment that would have and should have killed off all life that was trying to get a foothold. Also during this time, the Earth's crust cooled enough that rocks and continental plates began to form. The atmosphere would have been like a hot oven. The oceans, which condensed from the cooling atmosphere in the early part of the Achaean, would have been boiling hot. Radiation was immense. Conditions would have probably been less life-friendly than the internal environment of medical equipment that we now use all over the world today called sterilizers. To further complicate things, the moon was only 25,000 to 50,000 miles away from the Earth in those days. The tides produced by a moon that close would have been enormous; possibly hundreds of feet high. The swirling currents would have been unimaginably powerful.[4]

Abio-alchemists preach that enormous numbers of incredibly complex biochemical molecules somehow appeared then came together and gently assembled themselves in that hellish environment. The molecules somehow then brainlessly found their way into wandering micelles (fat bubbles) that had no purpose of existing other than lipid's **DL** hydrophilic and hydrophobic characteristics. Micelles stuffed with biochemicals then organized themselves into pre-biotic proto-cells. The proto-cells, which actually are non-existent, imaginary, and just part of this evo-illusion, for some very strange reason came alive, and then went on to become living bacteria on the early sea floor of that hellish Early Earth. Why did they come alive when the notion of life didn't exist within at least 30 trillion miles of Earth? Why didn't they just remain chemically stuffed fat bubbles? Why did they become chemically stuffed fat bubbles in the first place? Yes, and they did this all by themselves; oh, and with the help of random swirling **Dumb** Luck. Amazingly, many people do believe that notion and will argue it to the death.

How did the biochemicals that stuffed the micelles come to be? Of course evolution couldn't have been in the mix, because it didn't start until fully formed unicellular organisms with their DNA and genetic coding were in existence. Remember, according to evo-illusionists, evolution has nothing to do with evo-abiogenesis. Evolution occurs during cell division and the division of gametes, when DNA makes copies of itself. Those copy errors drive evolution, generation by generation. Without DNA and cell division there is no evolution. So evo-illusionists had to come up with an entirely new kind of evolution they call *chemical evolution*. Chemical evolution is defined as:

*The formation of complex organic molecules from simpler inorganic molecules through **chemical** reactions in the oceans during the early history of the Earth; the* *first step in the development of life on this planet. The period of chemical evolution lasted less than a billion years.*

Fig. 12-2 is a depiction of the Early Earth during the late Hadean or early Achaean eras, when nucleotides that self-formed and then sank to the bottom of the sea. Notice the size of the moon. This is what a moon only 25,000 miles away from Earth would look like. The swirling ocean currents and tides would have been enormous, as depicted. Can you imagine nucleotide molecules staying together sufficiently to form DNA or RNA in a sea like this? According to currently accepted scientific theory, nucleotides, the "zipper-teeth" in DNA and RNA, and amino acids formed in the atmosphere, then sank to the sea floor where there just happened to be, because of **Dumb** Luck, a type of clay called *montmorillonite clay*. Abio-alchemists teach that montmorillonite clay catalyzes, or aids the formation, of long chain molecules such as those that are precursors of RNA, DNA and proteins. So this clay nudged along the formation of long molecular chains. Those chains then chemically evolved, into precursors of proteins, RNA, and later DNA. Even in the evo-illusions concocted by evolution, it seems the Earth is on an intelligent track to produce life...and us. [5]

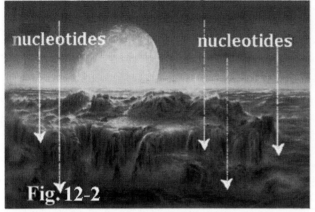

The only problem with montmorillonite clay being the catalyst for the formation of long chain biochemicals is that *phosphoramidate*, a very complex enzyme, is required along with the montmorillonite clay to catalyze the formation of these long chains. And phosphoramidate is not a biochemical that could have possibly existed the on sea floor of the Early Earth. For that matter, lipids (fats) that supposedly made up micelles (fat bubbles) could not have formed there either. Lipids are formed only by living organisms. Obviously there were no living organisms on that early sterile Earth. mRNA molecules, (the "one-sided zipper) supposedly are able to self catalyze their own replication. However when mRNA replicates, the molecules are then locked together like a closed zipper. More enzymes and life itself are required to unzip new mRNA copies. Since those enzymes didn't exist in the pre-biotic Earth environment, mRNA copies, even if they could have been brainlessly constructed without external assistance, would have been permanently locked to their progenitors. Utilization and further copying of the mRNA would have been impossible. Why would they have unzipped themselves anyway... unless they were knowingly preparing for the advent of life? [6]

Actually, proving evo-abiogenesis should be easy. All evo-alchemists need to do is take a sterile aquarium, fill it with sterile water, throw in some DNA ingredients like phosphate, ribose sugars, nucleotides, lipid cells, clay grains of their choice, and see if they can come up with DNA/mRNA wrapped in lipid micelles (fat bubbles). If cells of some kind don't form, evo-illusionists will of course say that is because it took millions of years. But the fat bubbles and DNA ingredients wouldn't last more than a few hours, weeks, or years at most. DNA deteriorates in water. Every policeman knows that. And, how long should one expect it would take for the DNA or mRNA to nestle inside of the micelle? A million years? Of course I'm being facetious. Evo-illusionists fool their audiences with their repeated "millions of years" fable. It should take no more than a few minutes, or hours; or at the worst, a few days. So c'mon abio-alchemists, demonstrate how mRNA and DNA form themselves, then move to the inside of micelles. In my mind I just can't get past the question of how do mRNA and DNA break the wall of the micelle to get inside? What exactly causes the opening that allows the mRNA and DNA to "swim" through. Abio-alchemists say mRNA clings to a clay particle, which then is surrounded by a fatty vesicle. So they should prove it. It can't take very long, since in an aquarium, the chemicals are all in close proximity and easily set up. This is the experiment they should be doing. They don't because they already know what the result would be. It would kill their *raison-d'etre*.

Still another problem for evo-illusionists is the fact that there are millions of dead animals (Fig. 12-3) all over the Earth at any one time. Many are composed of trillions of fully stuffed cells, ready to do life again. Their bodies form a perfect reservoir of all of the chemicals needed for life, including billions of DNA packets, carbohydrates, proteins, sugars, lipids… all in one big pile; all together. We shouldn't have to wait millions of years for mRNA or DNA stuffed micelles to form. These bodies, or reservoirs of all of the chemicals needed for life, if evo-abiogenesis is correct, should somehow form at least some new life; some new and living cells. These fantastic reservoirs of biochemicals and cell parts exist both on land and in water, in a much better environment than existed during the Hadean or Achaean. Biochemicals from dead animals should regroup like the biochemicals of evo-abiogenesis supposedly did long ago. At least some newly living cells should arise out of at least some of these "reservoirs". Dead animals containing all of the requirements for living cells should be billions of times more prone to form new life than any mix of Hadean fat bubbles and RNA ingredients swirling around in a hot ocean. Evolution shouldn't have to wait a billion years for a far inferior product. The entire Early Earth didn't have even one of these reservoirs of life's biochemicals. But, sadly for the world of abio-alchemy, no new life arises out of road kill. And if biochemicals from dead plants and animals can't regroup when they are in perfectly close proximity in a perfect life friendly environment, they sure didn't wiggle themselves together and form living cells in the hellish environment of the Early Earth.

Living cells are constantly balancing their chemistry and nutrition with their environment. The intake and outflow through the cell wall (osmosis) of ions, nutrients, and fluids maintain the cell's shape and size so that it doesn't collapse or explode. Cells act much like a water balloon with a pinhole leak. Liquid would have to be taken in as water goes out of the leak for the balloon to remain the same size. If water wasn't taken in, or was taken in too slowly while water was slowly exiting, the balloon would collapse. If water were taken in too rapidly while water was slowly exiting, the balloon would explode. Early proto-cells needed to evolve this balancing act. It couldn't have just come suddenly, according to evo-abiogenesis. Balance had to gradually form. The cell's internal control center had to absorb the task. All early proto-cells would have either collapsed or exploded before a full balancing system evolved; before full and constant balance was achieved. And that would have been the end of life on Earth.

To further complicate things for evo-abiogenesis, the fact that as a proto-cell assembled itself randomly and brainlessly, and came closer and closer to the ability to sustain life, it would be non-living or dead tissue. Dead cells are under a constant state of deterioration. *Necrosis* is the term given to describe the death of individual cells. Necrosis causes the cell to begin swelling. Material that makes up the nucleus is digested. Cell membranes break down. Cells that are studied by students and pathologists must be "fixed" with special chemicals to prevent that deterioration. The breakdown of a cell's biochemistry with cell necrosis will produce toxins that are damaging to nearby cells. Necrosis can actually lead to more necrosis. Cells that were struggling to come alive would have constantly been damaged by other nearby necrotic cells that didn't quite make it to life. The formation of the first living cell would have had to occur very isolated from other cells that are struggling to become alive.

Pre-biotic cells that were supposedly evolving into living cells would have constantly been fighting the fact that they were dead proto-cells. Could cells that are capable of supporting life fend off necrotic deterioration long enough to accept the "infusion" of life? Since life leaves a single-celled organism rapidly at death, is it thinkable that it would enter a single celled living organism slowly, however that event occurred? All living organisms, particularly single-celled types, are pretty much either living or dead. Different levels of "dead" only exist in nature in very short segments of time. So too, different levels of "infused life" must also have occurred in very short segments of time. If one small area of a cell were alive, the area of the cell that was not living would necrotize, and damage or kill off the small living area of the cell. Without living cells there could be no DNA. Without DNA, there could be no living cells. So what exactly was the source of the DNA that is the god-molecule and driving factor for evolution? Modern abio-alchemists have absolutely no idea. But that sure doesn't stop them from coming up with completely baseless absurd theories.

Abio-alchemists say that life was formed over a time span of hundreds of millions of years. This notion is what keeps believers in the fold. Anything can happen in hundreds of millions years. Anything. Time is one of the gods of evo-abiogenesis. In religion, God created everything. In evo-abiogenesis, Time is that god. Except Time did its creating in time spans that are so interminably long that they are incomprehensible to us mere mortals. And so Time has become a god for millions of people. Time certainly cannot be lab tested if the timespans are thousands to millions of years long, which is great for evolution. If it can't be tested, it cannot be disproved, and it certainly would be. In the world of evolution, all powerful and omnipotent Time resulted in the formation of mysterious entities, including ourselves, that we humans are incapable of comprehending. But it's obvious that this could not be the case, since "non-life to life" would have had to occur rapidly, just as "life to non-life" occurs rapidly. An entity coming to life could not have needed nor used billions of years to do so. Time would have been the enemy of cells trying to come to life, the killer, not the god or the friend.

Abio-alchemists tell how the Earth's atmosphere 3.8 billion years ago was full of hydrogen, hydrogen cyanide, methane, and ammonia, "among other" gases. They explain that the first major step to the goal of life is the combination of these four gases to form nucleotides, the "zipper-teeth" in DNA and RNA. But according to recent studies, "the original atmosphere was primarily helium and hydrogen". Heat from the still-molten crust, and the sun, plus a probably enhanced solar wind, dissipated this atmosphere. About 4.4 billion years ago, the surface had cooled enough to form a crust. It was still heavily populated with volcanoes, which released steam, carbon dioxide, and ammonia. This led to the Early Earth's "second atmosphere", which was primarily carbon dioxide and water vapor, with some nitrogen but virtually no oxygen. This second atmosphere had approximately 100 times more gas than the first atmosphere, But as it cooled much of the carbon dioxide was dissolved in the seas and precipitated out as carbonates. A later version of the "second atmosphere" contained largely nitrogen and carbon dioxide. However, simulations run at the University of Waterloo and University of Colorado in 2005 suggest that it may have had up to 40% hydrogen."[7-9]

So, the early atmosphere isn't quite what abio-alchemists describe, but since this is not an exact science, let's give them the benefit of the doubt and say the atmosphere was what they say it was and nucleotides could form. A first problem with the work of abio-alchemists is the fact that there were other gases circulating throughout the early atmosphere, not just the four needed to make nucleotides and life. The idea that just the four gases abio-alchemists have listed migrated together and formed nucleotides is ludicrous. The atmosphere and crust of the Early Earth was full of many kinds of gases, which were both reactants, and reagents, most of which had nothing to do with life. Many were deleterious to life.

The fact that we humans, intelligent as we are, cannot synthetically put together all of the parts of a cell and create life should give us a pretty good idea of what complete and utter brainless randomness in nature might be able to do, or not do. Without life in the form of living cells, there can be no DNA or nucleotides. The greatest intelligence that we humans are aware of, our own, cannot design, form, and assemble the parts of a cell, give it life, and form multi-celled synthetic organisms from synthetic unicellular organisms. If human intelligence cannot do this task, how on Earth could any intelligent person think a complete absence of intelligence in a caustic environment, as the Early Earth was, do the job? Can a complete lack of intelligence display incredible intelligence?

And still another problem for evo-abiogenesis: Researchers have pointed out difficulties for the abiogenic synthesis of nucleotides ("zipper-teeth") that form cytosine and uracil (two of the zipper-teeth that form DNA and mRNA). Cytosine has a half-life of 19 days at 100 °C and 17,000 years in freezing water. (A half-life is the amount of time that it would take 50% or half of any substance to decay or deteriorate. If you had a 1lb. pile of cytosine in a 100^0 C environment, in 19 days half would be destroyed. You would now have ½ pound of cytosine. In 19 more days you would have half of the half pound, or a ¼ pound. And on and on.)

Here are the conclusions of renowned evolutionists C. B. Thaxton, W. L. Bradley, and R. L. Olsen, in their book *The Mystery of Life's Origin*:

It has often been argued by analogy to water crystallizing to ice that simple monomers may polymerize into complex molecules such as protein and DNA. The analogy is clearly inappropriate, however... The atomic bonding forces draw water molecules into an orderly crystalline array when the thermal agitation (or entropy driving force) is made sufficiently small by lowering the temperature. Organic monomers such as amino acids resist combining at all at any temperature, however, much less in some orderly arrangement.[10]

In other words, chemical evolution, the **D**umb **L**uck formation of any bio-molecule, is an exercise in futility; or nothing but an evo-illusion.

Nobel Prize winning Belgian scientist Ilya Prigogine wrote:

The point is that in a non-isolated system there exists a possibility for formation of ordered, low-entropy structures at sufficiently low temperatures. This ordering principle is responsible for the appearance of ordered structures such as crystals as well as for the phenomena of phase transitions. Unfortunately this principle cannot explain the formation of biological structures. The probability that at ordinary temperatures a macroscopic number of molecules is assembled to give rise to the highly ordered structures and to the coordinated functions characterizing living organisms is vanishingly small.[11]

Maybe like $1:10^{190}$? A discussion of abio-alchemy and the earliest formation of DNA wouldn't be complete without discussing the two most famous abio-alchemists, Stanley L. Miller and Harold C. Urey. In 1953 these two abio-alchemists were thought to be on the verge of discovering the source of life on Earth. Working at the University of Chicago, Miller and Urey conducted an experiment that shook the world of evo-abiogenesis. Miller utilized methane (CH_4), ammonia (NH_3), hydrogen (H_2), and water (H_2O), chemicals, which at the time were thought to represent the major components of the atmosphere of the Early Earth nearly four billion years ago. They placed these chemicals into a closed system composed of specially designed flasks. (Fig. 12-4) Once Miller and Urey heated up the mixture sufficiently, they ran a continuous electric current through its vapor-filled chamber to simulate lightning storms believed to be common 4 billion years ago. In a week's time, Miller and Urey analyzed the contents of the flasks and found that a small portion of the chemicals had

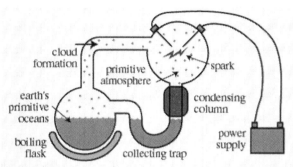

Fig. 12-4

reacted. A few amino acid molecules, the building blocks for proteins, which are made with DNA coding, had formed. They represented less than 3% of the total reactants, but they did get some amino acids.

Abio-alchemists, evo-illusionists, and evolutionauts the world over thought the Miller/Urey experiment proved that biochemical compounds could form on their own in the atmosphere and waters of the Early Earth. They were thrilled by these experiments. Miller/Urey had produced just what evo-abiogenesis needed to be a respected science: the first step of a pathway for the beginning of life through random processes. Did evo-abiogenesis finally have its replacement for spontaneous generation? As is usually the case with any form of evolution, the value of the Miller/Urey experiments were blown completely out of proportion. The making of amino acids in a lab flask by intelligent scientists who carefully selected the ingredients, placed them in an isolated flask, set up, controlled, and monitored the temperatures and pressures in the experiment only proves that *intelligence* was necessary to form the biochemicals needed for life. The first requirement in the Miller/Urey experiment was intelligence; putting very specific ingredients and conditions together. It was a perfect experiment that shows intelligence is necessary for the advent of life, and it proves **IID**. They oh so innocently had no notion of that fact. Their experiment was a fail for abio-alchemists, and a rousing success for those that think there is an amazing intelligence in nature.

A debate raged as to whether the Miller/Urey atmosphere truly represented an Early Earth atmosphere. Evidence seemed to show that it didn't. Then the pro-Miller/Urey scientists said that the Miller/Urey atmosphere could have existed near volcanoes, which were plentiful in those times. In reality, this debate is nothing more than a tempest in a teapot. It matters not. Forming a bit of scum composed of less than 3% amino acids is not within light years of forming DNA or even one simple protein molecule, much less a living cell. The amino acids were nothing more than impurities. One wonders how the 3% impurities that may have formed on the Early Earth got rid of the other 97% of the reactants that formed so proteins could make themselves in the next unlikely step. Miller/Urey demonstrated the uselessness of these types of experiments. Which is why they both posthumously qualify as abio-alchemists.

The Miller/Urey experiment required a constant and concentrated amount of electrical energy. While lightning storms were common fare on the Earth at the time of the first formation of life, they were not constant or concentrated like the "lightning" Miller and Urey used in their experiment. The high concentration and long-term administration of electricity and the selection and close proximity of the reagent chemicals didn't mimic any conditions on the Early Earth. Even so, Miller/Urey are frequently and staunchly referred to when evo-abiogenesis is discussed by evo-illusionists and abio-alchemists. They refer to Miller/Urey as if they formed living cells in the lab. A common statement in my debates with evolutionauts is: "Haven't you heard of the Miller/Urey experiments?" As if just the mention of Miller/Urey is enough of an argument all by itself.[12-14]

In reality, forming pure amino acid molecules is billions of light years away from the synthesis of even one protein molecule. The formation of proteins is billions of light years from forming living organisms. Miller and Urey weren't close to accomplishing either. Selection of ingredients and complete isolation of those ingredients from other chemicals requires intelligence.

I remember the first time I heard of the Miller/Urey experiments, I was one of those evolutionauts who was thrilled. My introduction to Miller/Urey came in a biology class in college. I *wanted* and *thirsted* for more information that would support my then new belief in the random formation of life and living species. Miller/Urey was it. Everything seemed to fit so perfectly. Chemicals in the atmosphere of the Early Earth were zapped by lightning, which formed amino acids, the building blocks for proteins and ultimately life. I could *visualize* that! Then unicelled organisms formed, followed by multicelled species; the phylogenetic tree gradually grew. At the other "modern" end of things, early primates evolved to become human beings! My skepticism and ability to give this science the objective perusal and independent thought that it needs was non-existent. I *wanted* to believe far more than I wanted to think out my newfound science on my own; which is the case with every evolutionaut I have discussed this subject with.

Miller and Urey became world famous because of their work. Evolutionauts were very optimistic that the puzzle of the formation of life would soon be solved. This has not been the case, however. Instead, the solution to this incredible *Puzzle, The Puzzle* of life's origins, seems only to have drifted farther and farther away from the grasp of science. The more scientists learn and discover, the farther away a solution gets; conversely, the closer abio-alchemists *say* it gets. And, unfortunately, the real solution, if there is one, is being blocked by abio-alchemists who are working to prove an impossibility; just like the early alchemists were also working on an impossibility.

So, here we are, dozens of years and thousands of experiments after Miller/Urey, and scientists are no closer to forming synthetic life in the lab or understanding the origin of DNA, RNA, proteins, and the other biochemicals of life than they were in 1953. Today we have the advantage of watching television science documentaries where abio-alchemists and evo-illusionists explain their experiments and research in great detail right before our eyes. Basic logic and rational objective thought are the slings and arrows that kill evo-abiogenesis and evolution. Belief is what keeps it going. With that being the case, I am amazed that there could be such faulty information and numerous fables so proudly displayed by evolution and evo-abiogenesis science documentaries. This science is so faulty, and so illogical, it should have been put in the trash bin of failed sciences long ago, where it belongs. But there it is, again and again, presented on television science documentaries over and over; the same astounding stuff, without the tiniest bit of skepticism from any of its highly respected participant speakers. No evo-illusionists or abio-alchemists on these documentaries ever question the information given by any of the other evo-illusionists or abio-alchemists as good scientists should. Without exception on the many evo-documentaries I have viewed, all of the evo-illusionists and abio-alchemists are in happy accord.

The hosts of these documentaries talk about evolution and evo-abiogenesis as if they're a lock. Everything happened the way they tell you, they're sure. "This evolved this way, that evolved that way, and these chemicals came together on the ocean floor near a volcano vent, and...." There simply is no question about it. They talk as if they're discussing the how's and why's of the manufacturing of an automobile. Most people know for sure how cars are brought into existence, and abio-alchemists and evo-illusionists put on airs of being just as certain about the random self-manufacture of living cells and their biochemicals such as DNA and RNA. They keep coming up with more and more evidence. Virtually all of their evidence goes against evo-abiogenesis, but they tout it as "proof" of their theory, whether their findings match evo-abiogenesis and evolution or not. They start out as if intelligence were not needed in their experiments. They then pour intelligence in as if it's one of the reagents. They then ignore the fact that they couldn't do the experiment without the input of intelligence. They hope, in fact, they know, they will fool the audience. They always do. They discuss evo-abiogenesis and evolution as if they were some sort of god, without the slightest notion that maybe things didn't happen the way they think and portray.

Chapter 13

Spontaneous Generation Just Won't Go Away

Never will the doctrine of spontaneous generation recover from the mortal blow struck by this simple experiment.- Louis Pasteur

It's really easy to observe how far abio-alchemists have come in the decades since 1953. The new version of Miller/Urey is evidence that this research has gone absolutely nowhere. But if the scientist that perform the experiments can fool the audience, and in particular, the politicians that hold the purse strings with OPM (other people's money), they are doing great. Science documentaries on stations such as PBS, the BBC, and Discovery are a great place to search out and study modern versions of spontaneous generation, AKA abiogenesis. Evo-illusion and abiogenesis documentaries all follow the same pattern, and will for a very long time. That pattern is a complete acceptance of evolution and chemical evolution without the slightest bit of skepticism. Before I demonstrate twenty-first century Miller/Urey, here is the formula for practically all evolution and evo-abiogenesis documentaries:

1. Respected evo-illusionists are the "stars" of evo-abiogenesis and evolution documentaries. They are typically very well educated and well known in the world of evolution; and usually very likable. Neil de Grasse Tyson and Bill Nye the Science Guy are great examples.

2. The "star" educates the viewers with absurd fantasies. In many cases the fantasies have been made up by the star evo-illusionist themselves. These shows wind up being promotions for their fantasy and in many cases promotions for a book they wrote on the subject.

3. All fellow evo-illusionists on the show agree with the "star" of the program, and symbolically pat him/her on the back.

4. No evo-illusionist or abio-alchemist ever questions or challenges a tale-telling evo-illusionist or abio-alchemist no matter how absurd the tale being told might be. Even if abiogenesis and evolution were truly the modus operandi for the formation of all of living nature, every tale that anyone makes up would not necessarily be how living nature originated. But any tale made up by one abio-alchemist or evo-illusionist is supported by all others. It's almost as if there's an unwritten rule: *No evo-illusionist shall challenge an evo-fable made up by another evo-illusionist, no matter how absurd the tale or circumstances may be.* I have yet to see an evo-illusionist even wonder aloud if any evo-tale *really* is correct. But I bet they do question quietly and to themselves, like Richard Dawkins obviously does.

5. Knowing smiles are freely given by evo-illusionists and evo-alchemists because they like to intimate that they now have immense knowledge that few people have, which solves a big part of *The Puzzle*, which is, in reality, currently unsolvable. They're going to relay this wonderful "solution" to the viewers.

6. Evo-documentaries give nearly all laboratory discoveries, new evo-fables, and fossil finds immense weight and celebration, even though most of the portrayed evidence doesn't deserve that weight and celebration, or it goes against what they're trying to prove.

The immense weight naively given "evidence" and information that "proves" evo-abiogenesis and evolution often results in a self-pat on the back, adulation from other evo-illusionists and abio-alchemists, and more grant money for their projects. Many evo-illusionists promote their books on these documentaries and wind up making a great deal of money. Why kill the goose that keeps laying the golden egg? Their mantra is, I pat you on the back, you pat me, and we both get adulation and big bucks.

These shows scientifically do nothing more than demonstrate how their abio-alchemists and evo-illusionists who most likely have been fooled themselves, are constantly trying to fool public viewers; mostly with great success. They are a perfect demonstration of the fooled in the act of fooling.

Here is a classic example of a modern Miller/Urey experiment. It comes packaged as *Nova Science Now*, a science series produced by Public Broadcasting Service (PBS). It supposedly demonstrates the "immense inroads" that have been made since Miller/Urey. Dr. John Sutherland and Dr. Jack Szostak, the "new" Miller/Urey duo, two renowned abio-alchemists, did experiments trying to solve *The Puzzle of the Beginning of Life*. I'll let you be the judge as to how far science has come in over a half of a century of searching for the origins of living organisms. The subject of this episode of Nova Science Now was *Where Did We Come From*. Dr. Neil DeGrasse Tyson, a very well known and charismatic astrophysicist, narrated it. I really like this guy when he discusses astronomy; obviously not when he discusses evolution and abio-alchemy. But this one is on the formation of life on Earth; way out of his normal realm. Unfortunately he lines up with all of the abio-alchemists on this show and he never questions them. He preaches abio-alchemy dogma as if it's pure scientific fact. This documentary is completely typical of evo-abiogenesis/evolution science documentaries and so is a great example of the dreadful science these have become.

Sutherland is from the University of Manchester. Szostak is a professor of genetics at Harvard Medical School. They were attempting to form nucleotides (zipper-teeth) in the lab, the building blocks for RNA and DNA, from their intelligently selected and fundamental ingredients. In doing so they thought they would be proving life came from non-living chemicals naturally; from **D**umb **L**uck happenstance on the Early Earth. There is no doubt abio-alchemists like Szostak and Sutherland are incredibly intelligent. Szostak received the Nobel Prize in Physiology. His contribution to the science of physiology is described in this way:

His discoveries helped clarify the events that lead to chromosomal recombination—the reshuffling of genes that occurs during meiosis—and the function of telomeres, the specialized DNA sequences at the tips of chromosomes. He is also credited with the construction of the world's first yeast artificial chromosome. That feat helped scientists to map the location of genes in mammals and to develop techniques for manipulating genes.[1]

An evolutionaut reading this book would certainly and expectedly ask, "How could Stephen Blume DDS, who is *just a dentist*, question a guy like Szostak. *C'est terrible!*" My reaction would be, how could someone with such great intelligence and knowledge actually think that the wonders he has studied in living cells put themselves together from complete randomness, and **Dumb Luck**? How could a person with a PhD, who has studied in incredible detail the wondrous designs of nature, and in particular the wonders of DNA I describe earlier in this book, come to the conclusion that there is no design; that **Dumb Luck** alone did the job of inventing and assembling nature's biochemistry?

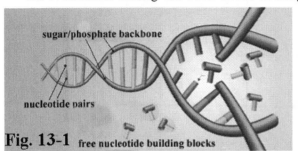

Fig. 13-1 free nucleotide building blocks

In this documentary John Sutherland said, "We are here on the planet, and we must be here because of organic chemicals", of course inferring that common organic chemicals just came together on their own and formed complex organic chemicals and eventually living cells, life, and us, with no other source entity involved.

The documentary goes on to introduce the *team* that was put together by Sutherland in an attempt to synthesize nucleotides, you will recall, the "T" shaped parts that make up DNA and RNA in the drawing. (Fig. 13-1) The *team* placed the ingredients for those nucleotides, ribose sugar and two half bases, in a flask, then "cooked" the flask and ingredients in a larger container of warm water. The larger container was supposed to represent a pond on the Early Earth. My first question is why was a "team" necessary to perform this experiment? Pouring a couple of reagents in a flask and getting reactants shouldn't require a *team*. There must be a logical reason, but it's hard for me to see. Do "teams" seem more scientific, and garner more grant money? There was no lab function on this show that could not have been easily performed by a single person.

The notion that many ponds, represented by the flasks being used, existed on the Early Earth doesn't even qualify as absurd. One would think tiny ponds would be far better than large bodies of water for the formation of any type of biochemical necessary for life. Currents and tides in large bodies of water would stir and dilute chemicals that needed to combine to allow evo-abiogenesis to take place. But small ponds, represented by the beakers in these experiments, would have evaporated quickly before any Earth shattering events such as the formation of RNA, DNA, and life could have taken place in them, nullifying what the audience is being shown anyway. Ponds and probably lakes and rivers would have formed and evaporated quickly, like a drop of water in a hot oven, in the heat of the Early Earth. And not one single pond out of billions would have had only the biochemicals used in this documentary. Sorry Sutherland and team.

Remember, evo-abiogenesis asserts that it took over a billion years to form life. Does anyone actually think a pond would last a billion years in the conditions of the Early Earth? Or could one last even a few weeks? And if this "experiment" supposedly represented what took place in the oceans of the Early Earth, just think of how diluted and spread out those chemicals would quickly become. Go to the beach and pour a couple of bottles of food coloring the in ocean, or any large body of water for that matter, if you want to see real diluting power in action. Or do it as a mind experiment and save the time. The results would be obvious.

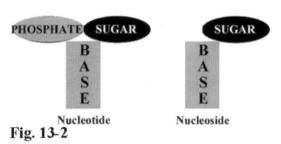

Fig. 13-2

At any rate, the recipe the team was using to form the precursor to nucleotides was ribose sugar (only a part of the top of the "T") and a base (the stem of the "T" or the "zipper tooth"). After many tries, the experiment did not yield the building blocks for RNA as they had hoped. Actually, phosphate groups are also in the formula for nucleotides. The ingredients for nucleotides should be: base + sugar + phosphate group. Phosphates were not mentioned by nor included in the team's recipe. The phosphate group makes up the rest of the top of the "T" along with ribose sugar. Did these scientists not include phosphate because they didn't want to confuse the audience? Or because they would have less chance of success with a phosphate group as an additional ingredient? What they were *actually* working toward was the building of what is called a nucleo*side*, not a nucleo*tide*. A nucleoside is a base + sugar with no phosphate group. (Fig. 13-2) If you don't understand what I have written about these ingredients, all you need to know is these are **not** the complete building blocks for nucleotides. Simply put, three different molecules are needed to make nucleotides. Let's call them A+B+C. Only A+B were used by the team. C was eliminated and not mentioned for obvious reasons. Making AB is far more likely and exponentially easier than making ABC.

The missing phosphate group really kills this experiment. It's a lot like Dawkins ignoring three quarters of a hemoglobin molecule in his evo-illusion. Let's ignore the missing phosphate groups and the fact that Sutherland is cheating like Dawkins, and get back to the laboratory procedures. No matter how the team cooked the ingredients, they came up empty handed. None of their experiments produced nucleotides, or even nucleosides. For some strange reason, someone got the idea of obtaining the help a very experienced chef. The idea was that "cooking" RNA building blocks might actually be somewhat like cooking *crème puffs*. I find the notion of experimenting with crème puffs very odd, but that was the scientific model used in this documentary. The team went to the kitchen of a very fancy local French restaurant. They tried their hand at baking crème puffs with the help of the chef and following the standard recipe. The results of their crème puff baking were awful. The crème puffs came out burned, flat, and not too appealing. The scientists had the same result as they did trying to bake nucleotides. They couldn't bake nucleotides or crème puffs.

Then the restaurant's head chef came to the rescue. He told them that to make crème puffs, they had to pre-cook the flour, crème, and sugar; everything except the eggs. Once the flour, sugar, and cream were precooked, *then* the eggs are added. All of the ingredients could be then be cooked together. The result would be beautiful crème puffs. The team tried it, and *voila!* They baked beautiful crème puffs. If you get nothing else out of this book, you will at least know how to make great crème puffs!

Fig. 13-3

The team then used the chef's technique to "cook" the ingredients for nucleotides. They precooked the ribose sugar with *half of a base* (half of a zipper tooth). They then added the other *half of the base*, and then cooked everything together. And guess what? *Voila encore*! They got a white scum on the bottom of the vile, which Sutherland said turned out to be nucleotides! At least that is what the audience is told, so I must *assume* that the scum at the bottom of the vile was truly nucleotides. (Actually nucleo*sides,* A+B but not A+B+C.) By placing a couple of the correct ingredients together, precooking those, then cooking everything together, they were able to fill in one of the roots of the Tree of Life! (Fig. 13-3) The tree roots are diagrammatical representations of the chemicals that went together to form the earliest living organisms. The trunk represents one celled life. The branches represent multicellular species. The tree in Fig. 13-3 was shown to the viewers to further fool them into thinking this was a valid and serious scientific experiment.

An immense problem with Sutherland's experiment is that ribose sugars synthesize in equally right-handed and left-handed molecules. It's the same problem abio-illusionists have explaining proteins with their all left-handed amino acids. The ribose sugars in DNA are all right-handed molecules. Imagine an asymmetric entity that is held in your hand and placed in front of a mirror. (see page 116) The entity in your hand would be the right-handed version; the image in the mirror would be the left. Or simply think of your right hand and left hand as molecules, which are mirror images of each other. It is not possible for nucleosides to have synthesized with only right-handed ribose; even in the crème puff oven. How were the lefties separated out and left behind? Again, like the missing phosphate groups, these abio-alchemists didn't mention a word about *chiral* molecules (right and left handed), or if their nucleosides came out with only the right-handed versions. They didn't mention right-handed ribose at all. Sutherland's team certainly didn't get results that formed nucleosides with only right-handed ribose. That's just part of the abio-illusion performed by these abio-illusionists. This also means, no matter what they do, they're wasting their time. Which is why it isn't mentioned in this documentary. Calculating the odds of an RNA molecule that is millions of base-pairs long randomly forming with only right-handed ribose molecules would be calculated by finding the product of $1:2^{millions}$; or two times itself millions of times. The *Ribose Number* would make the Hemoglobin Number look like chicken feed. This is information that is common knowledge for Sutherland and Szostak, but hidden from viewers of this documentary.

The inference from this documentary is that chemicals coming together, and nothing more, led to all of living nature. The "team" maintained their proud grins and acted as if they made an immense step towards conquering *The Puzzle* of the origin of life. They acted as if, after this immense pat-on-the-back-step, they had taken a giant leap toward synthetic life. Synthesizing living cells in the laboratory is "only ten years off"... "right around the corner", just as it was after Miller/Urey; just as it will be as long as evo-abiogenesis is the accepted scientific fable that describes the formation of the first living cells. Ninety-nine percent of the audience is not as savvy as you, the reader of this book. So ninety-nine percent of the audience will fall for this fake, and be all excited for days, thinking this is a huge step in the synthesizing of living cells.

After the team's successful formation of nucleo**sides** in beakers, Sutherland came up with a scenario on how nucleotides formed on the Early Earth. You see, the success of this minor experiment empowers Sutherland to tell humanity how nucleotides and ultimately cells formed from nothing. Using his own imagination, which is now a very illusory hyper-imagination, Sutherland said first there had to be a pond filled with ribose sugars and bases that bonded together. Again he doesn't mention that life uses only right-handed ribose molecules, or that no pond in the universe exists with just these ingredients; or that his ingredients require living organisms to exist. These two chemicals just happened to be in the same pond, for some very strange reason known only to Sutherland and his cosmic imagination, but the story is apparently accepted by the world of evo-abiogenesis. Next the ribose sugar + half bases evaporated into the atmosphere; and then, guess what. Yes, according to Sutherland's tale they condensed, and rained back down to Earth. They fell kind of like "organic snow". They somehow landed in *another* pond. And guess what was in the new pond-home for the ribose + base. Other chemicals that happen to make up nucleotides. Yes! By an incredible coincidence, and by Sutherland's imagination, they happen to be the biochemicals that the ribose sugar+ base were missing and dearly needed. They combined, and *voila*! Nucleotides! Do you believe this story? Sutherland certainly does. At least he says he does. So the ribose sugar+ base found their way into a second pond chock full with the other missing biochemical ingredients needed for nucleotides. And then, guess what. The biochemicals combined to form full nucleosides, or nucleotides, or actually anything Sutherland's imagination wants. It's his imagination we are studying here, nothing more. And we are well on our way to fully living cells! Well, there are a few trillion other steps and immense and impossible barriers that must be crossed first, but; at least we have the first step. Who am I to question the hard work and dedication of devoted and highly decorated abio-alchemists like Sutherland and Szostak?

I wonder if Sutherland read the results of experiments done to determine if ribose sugars could form all by themselves from randomness on the Early Earth. I think he missed this one, and many more: From a paper written on experiments done to determine the chance formation of ribose sugars and its longevity, Rosa Larralde*, Michael P. Robertson, and Stanley L. Miller, of Miller/Urey fame, from the Department of Chemistry and Biochemistry, University of California at San Diego, wrote:

The existence of the RNA world, in which RNA acted as a catalyst as well as an informational macromolecule, assumes a large prebiotic source of ribose or the existence of pre-RNA molecules with backbones different from ribose-phosphate. The generally accepted prebiotic synthesis of ribose, the formose reaction, yields numerous sugars without any selectivity. Even if there were a selective synthesis of ribose, there is still the problem of stability. Sugars are known to be unstable in strong acid or base, but there are few data for neutral solutions. Therefore, we have measured the rate of decomposition of ribose between pH 4 and pH 8 from 40°C to 120°C. The ribose half-lives are very short (73 min at pH 7.0 and 100°C and 44 years at pH 7.0 and 0°C). The other aldopentoses and aldohexoses have half-lives within an order of magnitude of these values, as do 2-deoxyribose, ribose 5-phosphate, and ribose 2,4 bisphosphate. These results suggest that the backbone of the first genetic material could not have contained ribose or other sugars because of their instability.[2]

So, you see, Stanley Miller himself throws ice water on the notion that chemicals could come together and form RNA. But this is certainly ignored by Sutherland and Szostak. They continue on as if their experiment has validity, when it has none. Remarkably, John Sutherland was given great accolades and awards for his and his team's work on discovering his own imaginary steps to life on the Early Earth. Here is a report from the Medical Research Council, Laboratory of Molecular Biology:

Fig.13-4

John Sutherland awarded the Royal Society of Chemistry Tilden Prize 2011

***John Sutherland** (Fig.13-4) has been awarded the Royal Society of Chemistry Tilden Prize for 2011. The Tilden Prize is awarded annually to recognise achievements in advancing the chemical sciences. It is awarded to those whose careers are defined by exceptional work, excellence and dedication. John's award is for his outstanding contributions to understanding the Origins of Life. A true understanding of biology must include knowledge of its chemical origin, and comprehension of the chemical events that gave biology its foundations. John's group, in the LMB's PNAC Division, is interested in uncovering prebiotically plausible syntheses of the informational, catalytic and compartment-forming molecules necessary for the emergence of life. At some stage an informational polymer must have arisen by purely chemical means and it has long been thought that this polymer was RNA... John's work has demonstrated the constitutional self-assembly of pyrimidine ribonucleotides from mixtures of simple building blocks - chemicals that existed on the Early Earth. John's group is now exploring similar 'systems chemistry' approaches to the purine ribonucleotides.*[3]

Sutherland's lab experimentation naturally led to the writing of a paper for Nature Magazine. Here is the abstract of that paper. If you are not educated in biochemistry, don't worry about reading it. It's such a perfect example of how abio-alchemists can get huge pats on the back for making mountains out of molehills, immense tales, nucleotides, and crème puffs.

Synthesis of activated pyrimidine ribonucleotides in prebiotically plausible conditions.

Powner MW, Gerland B, Sutherland JD. Nature. 2009 May 14;459(7244):239-42.

School of Chemistry, The University of Manchester, Oxford Road, Manchester M13 9PL, UK.

Abstract: At some stage in the origin of life, an informational polymer must have arisen by purely chemical means. According to one version of the 'RNA world' hypothesis this polymer was RNA, but attempts to provide experimental support for this have failed. In particular, although there has been some success demonstrating that 'activated' ribonucleotides can polymerize to form RNA, it is far from obvious how such ribonucleotides could have formed from their constituent parts (ribose and nucleobases). Ribose is difficult to form selectively, and the addition of nucleobases to ribose is inefficient in the case of purines and does not occur at all in the case of the canonical pyrimidines. Here we show that activated pyrimidine ribonucleotides can be formed in a short sequence that bypasses free ribose and the nucleobases, and instead proceeds through arabinose amino-oxazoline and anhydronucleoside intermediates. The starting materials for the synthesis-cyanamide, cyanoacetylene, glycolaldehyde, glyceraldehyde and inorganic phosphate-are plausible prebiotic feedstock molecules, and the conditions of the synthesis are consistent with potential early-Earth geochemical models.

I wonder if the chef that saved the day also was awarded Royal Society of Chemistry Tilden Prize? Gosh, without the crème puff chef, there wouldn't have been nucleosides. Doesn't he deserve the award as well? Maybe he got the Royal Society of Crème Puffs Award! Do I come across as a bit skeptical and droll? If I do, it's with good reason. How could intelligent scientists propose such nonsense with a straight face? Well, actually, their faces weren't straight. They have constant grins when discussing their Earth shaking laboratory experimentation. Maybe they think it's as absurd as I know it is, and they can't stop grinning, knowing they are able to fool so many people. For sure they fooled most people who watched this Nova Science Now documentary, and the people who put up the money to fund this scientific show of discovery as well. You see, to keep on doing what they are doing, they *must* come up with results, or the funds that pay for the team, crème puff chef, and cooking of biochemicals would be gone. Results are a necessity; an absolute must! They must get something. Anything. Possibly that is the cause of their grins. A good front will bring more good grant money. I wonder how much OPM (other people's money) is being spent on this type of experimentation. I'm sure lots.

Fig. 13-5

Continuing on with the miraculous discoveries of this documentary, Sutherland demonstrated what happens when they expose the nucleoside made of cytosine, one of the four bases (zipper-teeth) to light. Remember the white scum at the bottom of the number one beaker-pond, which was supposedly the cytosine nucleotide? Sutherland said, "Watch what happens when we expose the cytosine to light." His excited technician, a member of the *team*, turned on a purple looking light of some kind; a very scientific and expensive looking light. And, voila, right before our very eyes, the cytosine turned into guanine, another of the four bases in DNA! It was a real wow moment! Except for the fact that the white scum looked just like the white scum did before the heavy-duty light exposure. The Nova moderator was pretty excited, like he had just watched the first Wright brother's flight. Me? I didn't see anything happen. It was just white scum to white scum. No change. Maybe the white scum did change from cytosine to guanine just by being in the light. I would like to see the analytical data. This certainly makes one wonder about the longevity of cytosine. Does a few seconds of light turn all cytosine to guanine, which would destroy all cytosine; and all of life? I am sure that the scum was analyzed off camera, but, being the skeptic that I am... Sutherland made it seem like this was some sort of amazing scientific breakthrough. "You get two for the price of one!" said he.

My gosh, now mankind is twice as close to forming synthetic life! Billions of light years away, but twice as close. The team was so excited about their success that they went to a pub and had a few brews and knowing grins in celebration.

Having a laboratory technician place the exact ingredients needed for nucleotides in a beaker, then cooking the ingredients utilizing the cooking order provided by a crème puff chef, then finding that some of the molecules do come together to form nucleotides, isn't amazing. It's injecting intelligence when these abio-alchemists are trying to prove chemicals can come together on their own with no intelligence. Sutherland and his team completely blew it, and they have no idea. Sutherland's scenario could have never occurred on the Early Earth, or the modern Earth for that matter. Today, in a world filled with life and organic chemicals, do we see ponds with ribose sugar and half bases, and other ponds with just the needed missing half bases? Like matching puzzle pieces? Of course not. For that matter, do we see crème puffs forming on their own anywhere on Earth? The notion that the Early Earth had ponds and organic chemicals that performed like abio-alchemists Jack W. Szostak and John Sutherland tout they did is just an abio-alchemists dream; maybe a hallucination. These abio-alchemists are spending their lives on an obvious dead end. They're dedicating their lives to a fantasy, just like Charles Darwin was concerned that he himself was doing.

Working on abiogenesis in the lab, trying to make the chemicals of life, and trying to mimic conditions of the Early Earth is much like working on the construction of a space ship to Andromeda. Andromeda is a galaxy that is about 2.2 million light years from Earth. Remember the distance of a light year? Six trillion miles? Well, to make a space ship that could travel to Andromeda would require an engine that could accelerate the ship to 2.2 million times the speed of light…..or so. Of course that's impossible since nothing can travel faster than light speed. Szostak and Sutherland and the team are trying to do the equivalent of working on a rocket ship to Andromeda. They are doing the equal of seeing if they can make screws for the seats of that rocket ship. You see, they know there is an immense block wall that cannot be bypassed or penetrated. If they faced that fact, their projects would die. So would their income. So they must pretend their "step" is valid, and a good step in "the building of the rocket ship of life". What is their impenetrable wall? Actually there are thousands. But the largest two are the infusion of life into their chemicals, and the origin of the DNA code. I don't care how many chemicals they cook, and precook, how many chefs they have to help them, they will never ever form RNA or DNA that codes for proteins by filling beakers with carefully selected ingredients. Actually, they will never form any kind of living organism. They will never synthesize life. Their notion that random Early Earth chemicals eventually resulted in life couldn't possibly be more naïve. We humans don't even know what life is. Mankind hasn't even been able to come up with a definition of life. And the notion that someday we will build synthetic living cells, when we don't even know what life is, is beyond the pale. It's just another part of the DNA delusion. The same is true with putting together coded RNA or DNA. The code will never form in a beaker, even if there is a way to synthetically form RNA or DNA. These experiments are all equal to the engine of the Andromeda rocket ship. It can never be made by humans.

Good engineering and good science should realize that it must deal with the largest challenges first, before the lesser challenges are addressed. There are trillions of fully packed cells within easy reach of Sutherland and Szostak. What they need to do is bring non-living cells to life. If they cannot do that, it's silly to go through so many gyrations and billions of dollars to accomplish nothing. These abio-alchemists celebrated and toasted a great victory, which it wasn't. It's not worth a beer at the pub. The only good outcome of this experiment is that the team can now make good crème puffs.

Sutherland said, of his team's experiment, "My team and I have recreated an Early Earth scenario and let it run. The chemistry just does it on its own." He proved **IID**, and proved that random chemicals cannot and did not come together on their own and form the biochemicals required for life.

Sutherland said: "You actually have to be the person that writes the recipe book." Right. Doesn't he realize that it takes intelligence to write recipe books? Everything he says, everything he does proves the need for intelligence in the designs of living nature. Sutherland so innocently has no clue. He goes on and on discussing events and entities that require intelligence for their origination, and he has no idea. Why should scientists that support intelligent design waste their time and money doing any experiments? Experiments that do nothing but support the **IID** position are constantly being performed for them by abio-alchemists like Sutherland and Szostak, and Miller and Urey. And just think. American and British taxpayers are paying through the nose for these worthless experiments.

The Nova narrator continued: "It came together in simple steps that could have happened on their own on the Early Earth. To cook these chemicals, you need 14,700 pounds of pressure and temperature of 212^0 to 320^0 F. Where in the world do your find a kitchen like that? Volcanic vents thousands of feet below the surface here on Earth."

So now we need not only big and little ponds, but also very deep, hot ponds, and immense pressure! And, the moderator continues, these imaginary processes "could have happened on their very own?" In reality, the team proves that it *cannot* happen on its very own. For life's biochemicals to form, a team and probably a crème puff chef are an absolute necessity. If cells formed on their own near volcanic vents where the pressure was thousands of pounds, and the temperature very very hot, why don't we see those biochemicals forming today in a world that is full of biochemical building blocks and hot volcanic vents? These are the same hot volcanic vents with the *exact same* conditions as were present on the Early Earth. Why don't we test the theory and send a diving bell down to these vents and see if biochemicals really are forming. The answer: because organic chemicals are not forming. Doing that would kill these simplistic experiments and the grants that go with them. Actually there is a big difference between the volcanic vents on the Early Earth and the modern versions. Modern oceans are loaded with biochemicals. The tides and currents today are much more amenable to life. The moon is 250,000 miles away, instead of 25,000 to 50,000 miles away as it was 3.8 billion years ago. The likelihood of finding randomly forming biochemicals on a modern Earth should be billions of times greater if Sutherland and Szostak are right.

Finishing the documentary, Szostak said: "Life emerged from chemicals. Then, after that, it's just the details." Here is a PhD biologist who knows all about the incredibly complex hyper-drive miniature-machines that cells are. And he thinks "it's just the details"? A truly amazing statement. I think there is more to it than that, Jack. Szostak again: "It never occurred to me to put them together in different order, so it's sure not obvious." Not obvious, but an Early Earth without a lick of intelligence was able to do it? "Not obvious" which actually proves the team's and a chef's intelligence was the most important ingredient needed to figure out how to cook just three chemicals. Intelligence is the most important ingredient you need Jack. You proved **Ingenious Invention and Design (IID)** and you have no idea. By the way, "not obvious" is one of the major criteria for an invention to be patentable at the US Patent Office.

Szostak again: "For early life, you need two things: a cell membrane and you need genetic material. Something that can allow the inheritance of information". Again, Szostak, Mr. Nobel Prize winner, needs one more very important ingredient that he forgot about: the ability of the team to bring those non-living membranes and genetic material to life. Yes Jack, life. You can have all the cell membranes and genetic material you want. But "life" will forever be your Andromeda rocket ship; which is why you don't bring non-living fully packed cells with all biochemical back to life as a good scientist would and should. This experiment is nothing but a bait and switch exercise, which obviously fools most of the audience. If fully formed and loaded cells that just died cannot be brought back to life, the pitiful few biochemicals these intelligent lab techs force together are nothing but an incredible waste of time. But these experiments are evo-illusions with a cause. The cause is the fooling of the audience, the pats on the back that they give each other, and the bringing in of more and more funds so more worthless projects like this one can be performed.

So what is my stance on this greatest *Puzzle* in the history of mankind? It is completely obvious that abiogenesis occurred. My argument isn't that there was no abiogenesis. My argument is, exactly how did it take place? Evolutionauts and religious creationists would both agree that abiogenesis occurred. At one time in the history of the Earth there was no life; then there was. But, exactly how did life begin? Neither evo-abiogenesis nor abiogenesis of the religious type have ever been observed in nature or displayed in any laboratory experiments. The only rational scientific answer is we humans simply don't know. And we aren't close to figuring out this immense *Puzzle*. Real science should be willing to admit that fact instead of coming up with an immense pile of fantasies, fake documentaries, and useless experiments.

To hopefully gain a new perspective on abio-alchemy and evo-abiogenesis, I read a book titled: *What Is Life? Investigating the Nature of Life in the Age of Synthetic Biology* by Ed Regis, published by Ferrar, Straus, and Giroux, New York 2008. In this book the writer describes several consortiums and their laboratory operations that were put together to try to synthesize a living cell. Anyone with a little knowledge and a minimum amount of common sense like any reader of this book could figure out that this is a giant waste of time and money. One group of gullible investors had donated over $14,000,000 to fund these projects; as if lots of money will eventually solve *The Puzzle*. The wishful thinkers and donators were made up of naïve and inadequately educated believers and government agencies. Unfortunately, the government agency took money for this useless project from innocent taxpayers, as always without their permission of course. After blowing the millions of dollars, the obvious conclusion was that living cells cannot and never will be synthesized; that we are not even remotely close to being able to synthesize life. In the end the writer answers the question posed by the cover of his book. If one insists on a scientific definition for *life*, Regis suggests the following: *Defining life as embodied metabolism seems to be the most defensible theory we have at the present.* That's it? *Embodied metabolism*? After blowing tens of millions of dollars in endeavors similar to those described earlier in this chapter, and in numerous other attempts at synthesizing life, that's all they have? Regis says the consortium and study left the search for the origin of life and the synthesis of living cells exactly where they started: nowhere. He admits spending tens of millions of dollars produced no progress in determining the source of life, and in synthesizing living cells. What was Regis final comment? *"And that's life."*

Reading *What Is Life?* gave me no more incite as to what life really is, which was as expected. No one really knows or has that answer. And if we don't even know what life is, can it ever be synthesized? Will we ever synthesize a living cell? Will we ever truly understand what directs the morphing of a fertilized ovum into an embryo and then into a fetus and then a fully formed infant... and then into an adult human? Will we ever figure out what directs the location and functions of the nearly infinite number of molecules inside the cells in the human body? Will we ever comprehend what directs all of the cells in our bodies to travel to the locations where they are needed, and what directs their functions? We have come so far toward understanding *what* our cells and their internal contents do. In doing so, we have created greater mysteries than we could ever have imagined; mysteries that were once thought solvable that probably never will be. In the 1800's and before, we had no idea what we didn't know. Now we do. In a sense, since the advent of modern scientific study and experimentation, we have moved light years backward from where we thought we were in truly understanding life. While we know so much more *what*, we know so much less *how*...and *why*.

Index:

abio-alchemist 119-140
abiogenesis 56, 120-142
alternative splicing 46,47
Andromeda Galaxy 139-142
Angelman syndrome 57
antiparallel 31
Achaean 126
Atmosphere 121-128, 135
Avery, Oswald 23
axon terminals 43
Bacon, Francis 60
bacterial flagellum 74
base pair 23, 28, 31, 32, 44, 64, 68, 69, 82, 83, 84, 85, 95
bases 22, 28, 29, 31, 32, 44, 63, 67, 68, 83, 132, 135, 137, 138
bat sonar 108
bent-winged bat 40
binary code 60, 61
binary fission 82
blastocyst 71, 76
blood clotting 88
Blume's Law 55, 74, 75
Blume's Theory of Life 105
Boveri, Theodor 20, 21
Campbell, Keith 71
Cavendish Laboratory 26
cell division 22, 28, 29, 31, 32, 44, 67-69, 73, 80, 82, 84, 103-105, 121
cell membrane 58, 73, 87-90, 103, 124, 140
cell theory 15
characteristics 8, 15, 18, 20, 21, 22, 33, 37, 41, 42, 51, 54, 57, 81, 103, 104, 121
Chargaff, Erwin 23, 24
Chargaff's Rules 24
chemical evolution 121, 122, 126, 130
Chirality 115, 134
chromatin 20-22
chromosomal theory of inheritance 22
codon 28, 44, 45, 62, 65, 112, 113
Crick, Francis 7, 16, 24, 25, 26-37
cytoplasm 58, 74, 75, 93, 94
cytosine 20, 23, 24, 28, 31, 126, 138
Darwin, Charles 53, 54, 55, 102, 138
Dawkins, Richard 3, 5, 6, 13, 34, 35, 41, 48, 52- 54, 96-102, 107-118, 130, 133

dendrite 43
deoxyribonucleic acid 7
differentiation 71, 76, 77, 94, 99
diploid 72
Discover Magazine 14
DNA Learning Center 88
Dolly the sheep 72-73
Dumb Luck (DL) 35, 37, 51, 58, 75, 107-132
dwarf stars 49
effector protein 89-90
Einstein, Albert 49
electromagnetic force 50-52
enzyme 42, 56, 65-67, 79-81, 102, 122
epigenesis 99-102
Escherichia coli 82-84
essential amino acids 59
eukaryote 79, 84
Evo-illusion 11
Evo-illusion of Man 11, 49
external cellular domain 76, 87-99, 105
fetal development 77, 101, 104
Flemming, Walter 20
Franklin, Rosiland 26-34
fruit flies 21, 72
Gamow, George 27-29
Gemmules 15
genes 38-46, 46, 47, 54, 57, 59, 67, 72, 74, 77, 79, 87, 93-103
Genome Project 39-43, 69, 72, 78, 85, 93,
glial cells 43-44
gravitational waves 49-50
gravity 49-52
green puffer fish 40
growth hormone 1 and 2 57
Hadean 120
Half-life 126
haploid 72
Hebrew University of Jerusalem 94
helicase 67
hemoglobin number 110-118
homunculus 95
Howard Hughes Medical Institute 42
HOX genes 14, 46, 77
ID, intelligent Design 36, 37, 139
IID, Ingenious Invention and Design 36, 37, 100, 112-116, 127, 139, 140
information 131-140
input proteins 89

internal cellular domain   87-89
invention   35-37, 51, 52, 58, 60, 81, 116, 117, 140
junk DNA   44, 46
King's College Laboratory   29, 32-34
Kossel, Albrecht   20
Krebs cycle   69
Leibniz, Gottfried   60
Levene, P. A.   22-24
life   35, 36, 42, 51-53, 55, 141
lipid   58, 120-123
Lipton, Bruce   73
lysosome   58
marbled lungfish   40
meiosis   22, 80, 131
melanin   57
Mendel, Gregor   18, 20, 21
micelle   120, 123
microtubule filaments   69, 75
Miller, Stanley L.   126-131, , 136, 140
mitochondria   58, 73
mitosis   20, 67
montmorillonite clay   122
Moore, Gordon E.   54, 62
Moore's Law   54
Morgan, Thomas   21
morphogens   75-77
Morse code   38, 59, 60
Morse, Samuel F. B.   59
motor proteins   75
mRNA   29, 44, 66, 67, 81, 122, 123, 126
mutation   14, 15, 22, 38, 39, 46, 47, 77-84, 101, 102, 116
National Institute of Health   41
necrosis   124
neuron   43-44, 104
neurotransmitters   43
neutron stars   49
Nobel Prize   20, 25, 27, 34, 55, 105, 126, 131, 140
Nova Science Now   131, 137, 138, 140
nuclear transfer   71
nucleic acid   20, 24, 33,
nuclein   19, 20
nucleoside   133-138
nucleotides   13, 312, 40, 64-68, 79, 96, 122, 123, 125, 126, 131-138
Nye, Bill   102, 103
organelle   58, 67, 73, 84, 93, 101, 104

Origin of Species by Means of Natural Selection   3, 53
osmosis   124
output proteins   89
ovists   96
pangenesis   15
Pauling, Linus   25, 27, 30, 34
PAX homeobox genes   77
phospholipids   58
phosphoramidate   122
Pray, Leslie A.   79-84
preformation   95, 98
prokaryote   83-84
protein   13-14, 16, 19, 22, 28, 31, 32, 33, 35, 36, 38, 40, 42, 44-47, 56-59, 62-67, 70, 73, 74, 75, 77, 81-86, 89-93, 95, 97, 101-103, 108, 109, 111, 112, 114-118, 122, 123, 126-129, 134, 139
proto-cell   121-124
proton   51
receptor protein   89, 90
red blood cell   73, 74, 113, 114
red viscacha rat   40
Regis, Ed   141
replication   67-69, 79, 82, 85, 103, 122
reticulocyte   73
ribosomes   9, 44, 57, 58, 67, 81
RNA polymerase   65, 66
RNA Tie Club   28
S. aureueus   83
Schleiden, Matthias   15
Schwann, Theodore   15
singularity   51
Smith, Francis   60
Sutherland, John   131-139
spermists   96
spontaneous generation   119-120, 127, 130
starlings   100, 101
Stated Clearly   91, 92
strong force   50-52
supernovae explosions   49
Sutton, Walter   20, 22
Szostak, Jack   131-140
tetranucleotide hypolthesis   23
Thaxton, C. B.   126
The Blind Watchmaker   3, 107, 110, 115
The God Delusion   13, 46, 48, 96
The Greatest Show on Earth   97, 98

The Mystery of Life's Origin 126
The Selfish Gene 96, 99
Thor the Thunderer 48, 49
Time Magazine 39, 98, 99
traits 8, 15, 18, 21, 22-24, 33, 37, 41, 42, 103,
transistors 54, 61, 62
tree of life 134
Tyson, Neil de Grasse 130, 131
ubiquitin protein ligase E3A gene 57
University of Colorado 125
uracil 20, 126
Urey, Harold 126-140

Vortisick, Frank. T. 93
Watson, James 7, 8, 16, 24, 25, 26-36, 44, 53, 54, 77, 84, 90
weak force 50, 52
What Is Life 141
Where Did We Come From 131
Wilkins, Maurice 33, 34
X-ray studies 16, 26, 29, 30, 31, 33
XX and XY chromosomes 102
Zeus 48, 49, 69, 93

References:

Chapter 1

1. Discover Magazine, November, 1992, by Mark Caldwell
2. HOX GENES: Seductive Science, Mysterious Mechanisms, Ulster Medical Society, Terrence R Lappin, et al, January, 2006
3. Understanding Evolution, http://evolution.berkeley.edu/evolibrary/article/hoxgenes_01
4. Biography, Ernest Haeckel (1834-1919) http://www.ucmp.berkeley.edu/history/haeckel.html
5. Charles Darwin's Theory of Pangenesis. Zou, Yawen, *Embryo Project Encyclopedia* (2014-07-20). ISSN: 1940-5030 http://embryo.asu.edu/handle/10776/8041

Chapter 2

1. Gregor Mendel: a Biography http://www.biography.com/people/gregor-mendel-39282
2. DNA and Proteins are Key Molecules of the Cell Nucleus, DNA From The Beginning, http://www.dnaftb.org/15/bio.html
3. http://simplyknowledge.com/popular/biography/friedrich-miescher
4. Famous Scholars from Keil: http://www.uni-kiel.de/grosse-forscher/index.php?nid=flemming&lang=e
5. The Nobel Prize in Physiology or Medicine: https://www.nobelprize.org/nobel_prizes/medicine/laureates/1910/kossel-bio.html
6. Theodor Heinrich Boveri (1862-1915) By Inbar Maayan, published: 2011-03-03 https://embryo.asu.edu/pages/theodor-heinrich-boveri-1862-1915
7. https://www.britannica.com/biography/Walter-Sutton
8. Pheobus Levene, 1869-1940 http://www.dnaftb.org/15/bio-2.html
9. Oswald Avery, DNA, and the transformation of biology by Matthew Cobb http://www.cell.com/current-biology/pdf/S0960-9822(13)01515-7.pdf
10. Edwin Chargaff, biography, by Eugene M. McCarthy, PhD http://www.macroevolution.net/erwin-chargaff.html

Chapter 3

1. Biography: James Watson https://www.biography.com/people/james-d-watson-9525139
2. Biography: Francis Crick, https://www.biography.com/people/francis-crick-9261484
3. Chemical Heritage Foundation: James Watson, Francis Crick, Maurice Wilkins, and Rosiland Franklin
4. https://www.chemheritage.org/historical-profile/james-watson-francis-crick-maurice-wilkins-and-rosalind-franklin

5. Wikipedia, Francis Crick: https://en.wikipedia.org/wiki/Francis_Crick
6. Wikipedia, James Watson: https://en.wikipedia.org/wiki/James_Watson
7. The Physics of the Universe, George Gamow, (1904-1968) http://www.physicsoftheuniverse.com/scientists_gamow.html
8. U.S. National Library of Medicine, The Francis Crick Papers, The Discovery of the Double Helix, 1951-1953, https://profiles.nlm.nih.gov/SC/Views/Exhibit/narrative/doublehelix.html
9. Medical Research Council, Insight, Behind the Picture: Photo 51 https://www.insight.mrc.ac.uk/2013/04/25/behind-the-picture-photo-51/
10. The Pauling Blog, The X-ray Crystallography that Propelled the Race for DNA, https://paulingblog.wordpress.com/2009/07/09/the-x-ray-crystallography-that-propelled-the-race-for-dna-astburys-pictures-vs-franklins-photo-51/
11. Maurice Wilkins, Behind the Scenes of DNA, SciTable by Nature Education, https://www.nature.com/scitable/topicpage/maurice-wilkins-behind-the-scenes-of-dna-6540179
12. Independent: Anger at Slur at Rosiland Franklin's Colleague, http://www.independent.co.uk/news/people/photograph-51-anger-at-slur-on-rosalind-franklins-colleague-in-new-play-about-the-dna-pioneers-a6668466.html
13. Richard Dawkins Quotes: https://www.goodreads.com/quotes/83303-we-are-going-to-die-and-that-makes-us-the

Chapter 4:

1. National Human Genome Research Institute: https://www.genome.gov/10001772/all-about-the--human-genome-project-hgp/
2. Scientific American, Wojciech Makalowski, What is Junk DNA, and What is it Worth? https://www.scientificamerican.com/article/what-is-junk-dna-and-what/
3. National Institute of Health National Human Genome Research Institute: https://www.genome.gov/10001772/all-about-the--human-genome-project-hgp/
4. Genetic Changes Shaping the Human Brain, Howard Hughes Medical Institute, by Byoung-il Bae, Christopher Walsh, et al, Feb. 23, 2015 https://www.ncbi.nlm.nih.gov/pmc/articles/PMC4429600/
5. The Human Brain in Numbers: A Linearly Scaled-up Primate Brain, Frontiers in Human Neuroscience, Suzana Herculano-Houzel, Nov. 9, 2009, https://www.ncbi.nlm.nih.gov/pmc/articles/PMC2776484/
6. https://blogs.scientificamerican.com/brainwaves/know-your-neurons-what-is-the-ratio-of-glia-to-neurons-in-the-brain/
7. http://blog.eyewire.org/how-do-we-know-that-there-are-100-billion-neurons-in-the-brain/
8. http://hyperphysics.phy-astr.gsu.edu/hbase/Forces/funfor.html
9. Scitable by Nature, transfer RNA/tRNA, https://www.nature.com/scitable/definition/trna-transfer-rna-256

Chapter 5:

1. https://www.decodedscience.org/comparing-genetic-code-dna-binary-code/55476
2. Alternative RNA Splicing: http://www.exonhit.com/technology/alternative-rna-splicing
3. Nature Reviews, Alternative splicing: a pivotal step between eukaryotic transcription and translation: http://www.nature.com/nrm/journal/v14/n3/full/nrm3525.html
4. Mechanisms of alternative splicing regulation: insights from molecular and genomics approaches: https://www.ncbi.nlm.nih.gov/pmc/articles/PMC2958924/
5. What Are Gravitational Waves, https://www.ligo.caltech.edu/page/what-are-gw
6. What Are Gravitational Waves and Why Do They Matter: Popular Science, http://www.popsci.com/whats-so-important-about-gravitational-waves
7. Six Things You Probably Didn't Know About Gravitational Waves: https://www.wired.com/2016/02/6-things-you-probably-didnt-know-about-gravitational-waves/
8. A Briefer History of Time, The Forces of Nature, and the Unification of Physics: https://brieferhistoryoftime.com/chapters/11
9. Fundamental Interaction: https://en.wikipedia.org/wiki/Fundamental_interaction
10. 50 Years of Moore's Law: http://www.intel.com/content/www/us/en/silicon-innovations/moores-law-technology.html
11. http://anthro.palomar.edu/blood/blood_components.htm
12. NCBI: https://www.ncbi.nlm.nih.gov/books/NBK2263/
13. SciTable by Nature Education: https://www.nature.com/scitable/topicpage/eukaryotic-cells-14023963
14. Cell Structure and Functions: http://biologicalphysics.iop.org/cws/article/lectures/52854

Chapter 6

1. http://aminoacidinformation.com/how-many-amino-acids/
2. Biography, Samuel F. B. Morse, https://www.biography.com/people/samuel-morse
3. Bacon's Bilateral Cipher: http://www.math.cornell.edu/~morris/135/Bacon.pdf
4. Gottfried Wilhelm Leibniz https://en.wikipedia.org/wiki/Gottfried_Wilhelm_Leibniz
5. Gottfried Wilhelm von Leibniz: http://www-history.mcs.st-andrews.ac.uk/Biographies/Leibniz.html
6. Claude Shannon, a Goliath Among Giants://www.bell-labs.com/claude-shannon/
7. What is Protein Synthesis?: http://www.proteinsynthesis.org/what-is-protein-synthesis/

8. Protein Synthesis:
http://chemistry.elmhurst.edu/vchembook/584proteinsyn.html
9. SciTable by Nature: https://www.nature.com/scitable/topicpage/ribosomes-transcription-and-translation-14120660
10. Your Game: What is Mitosis: https://www.yourgenome.org/facts/what-is-mitosis
11. The Biology Project: The Cell Cycle & Mitosis:
http://www.biology.arizona.edu/cell_bio/tutorials/cell_cycle/cells3.html

Chapter 7

1. Animal Research Project, Cloning Dolly the Sheep:
http://www.animalresearch.info/en/medical-advances/timeline/cloning-dolly-the-sheep/
2. 20 Years After Dolly the Sheep Led the Way-Where is Cloning Now? Scientific American, Karen Weintraub, July 5, 2016
https://www.scientificamerican.com/article/20-years-after-dolly-the-sheep-led-the-way-where-is-cloning-now/
3. University of Edinburgh, The Life of Dolly,
http://dolly.roslin.ed.ac.uk/facts/the-life-of-dolly/index.html
4. Worm Genome Sequencing Influenced Human Genome Project's Dada Sharing Principles, Kathie Y. Sun, NHGRI Scientific Program ,
https://www.genome.gov/27552817/worm-genome-sequencing-influenced-human-genome-projects-data-sharing-principles/
5. You Share 70% of Your Genes With This Slimy Marine Worm, Live Science, Mindy Weisberger, November 18, 2015
https://www.livescience.com/52843-acorn-worm-genome-sequencing.html
6. Reticulocyte, https://en.wikipedia.org/wiki/Reticulocyte
7. The Biology of Belief p. 65, Mountain of Love/Elite Books, March, 2005, Bruce Lipton
8. Cytoskeletons and Motor Proteins, https://www.tocris.com/cell-biology/cytoskeleton-and-motor-proteins
9. Motor Proteins and Molecular Motors: How to Operate Machines at Nanoscale, Anatoly B. Kolomeisky, Oct. 7, 3013:
https://www.ncbi.nlm.nih.gov/pmc/articles/PMC3858839/
10. Morphogen Gradient Formation, Ortrud Warlick, et al, September, 2009
https://www.ncbi.nlm.nih.gov/pmc/articles/PMC2773637/
11. PubMed, Morphogens and cell survival during development, Mehlen, P. et al, Sept. 15, 2005, https://www.ncbi.nlm.nih.gov/pubmed/16041752
12. Morphogen gradients in Development: from form to function, Jan L. Christian, Nov. 11, 2011,
https://www.ncbi.nlm.nih.gov/pmc/articles/PMC3957335/
13. Development, Morphogens, their identification and regulation, Tetsuya Tabata, et al, http://dev.biologists.org/content/131/4/703
14. U.S. National Library of Medicine, Homeoboxes,
https://ghr.nlm.nih.gov/primer/genefamily/homeoboxes
15. Pediatric Research, Homeobox Genes in Embryogenesis and Pathogenesis, Manual Mark et al, 1997,
https://www.nature.com/pr/journal/v42/n4/full/pr19972506a.html

Chapter 8

1. Science Daily, Bacteria Mutate Much More Than Previously Thought, Aug. 9,2017
https://www.sciencedaily.com/releases/2007/08/070818112338.htm
2. SciTable, by Nature; Antibiotic Resistance, Mutation Rates and MRSA, by Leslie Pray, PhD, 2008 https://www.nature.com/scitable/topicpage/antibiotic-resistance-mutation-rates-and-mrsa-28360
3. Genes and Development, CSH press, Protein-protein interactions define specificity in signal transduction, Pony Pawson and Piers Nash, http://genesdev.cshlp.org/content/14/9/1027.full.html&sa=U&ei=IuDGVPeDF87joASBioCgCQ&ved=0CEwQFjAJ&usg=AFQjCNHza6Uw9NlDaA5Ogcg5rMylhRi8Vw
4. The Biology of Belief p. 65, Mountain of Love/Elite Books, March, 2005, Bruce Lipton

Chapter 9

1. Advanced Tissue: https://www.advancedtissue.com/the-6-steps-of-the-wound-healing-process/
2. The DNA Learning Center: https://www.dnalc.org/view/16877-Cell-Signals.html
3. Virtual Cell Textbook:
https://www.ibiblio.org/virtualcell/textbook/chapter3/cmf1c.htm
4. Boundless: Types of Receptors,
https://www.boundless.com/biology/textbooks/boundless-biology-textbook/cell-communication-9/signaling-molecules-and-cellular-receptors-83/types-of-receptors-381-11607/

Chapter 10.

1. https://www.patreon.com/statedclearly
2. What is DNA and How Does it Work?
https://www.youtube.com/watch?v=zwibgNGe4aY&t=89s
3. The Genius Within, by Frank T. Vertosick, Jr. Harcourt Books, 2002, pp. 116-120
4. Mechanism That Triggers Differentiation of Embryo Cells Discovered: https://www.sciencedaily.com/releases/2008/12/081221220328.htm
5. Preformation and the Enlightenment:
https://embryo.asu.edu/pages/preformationism-enlightenment

Chapter 11

1. Chirality, Wikipedia, https://en.wikipedia.org/wiki/Chirality_(chemistry)
2. Stereochemistry: Determining Molecular Chirality, UCLA, http://www.chem.ucla.edu/~harding/tutorials/stereochem/id_mole_chiral.htm

3. PubMed: Hydrogen bonds in crystal structures of amino acids, peptides and related molecules.
https://www.ncbi.nlm.nih.gov/pubmed/521211
4. Boundless, Amino Acids
https://www.boundless.com/biology/textbooks/boundless-biology-textbook/biological-macromolecules-3/proteins-56/amino-acids-303-11436/
5. Proteins: Not Just a Chain:
Hemoglobin, http://en.wikipedia.org/wiki/Hemoglobin
6. Report of National Center for Science Education, vol. 20, no.4, 2001. David H Bailey http://www.dhbailey.com/papers/dhb-probability.pdf
7. Bentley WA, Humphreys WJ. *Snow Crystals*. NY: Dover Publications, 1962.
8. Foster D. *The Philosophical Scientists* New York: Barnes & Noble Books, 1993.
9. Hoyle F., Wickramasinghe, C. *Evolution from Space* London: JM. Dent and Sons, 1981.

Chapter 12

1. Micelle Formation: Arne Thomas
https://www.mpikg.mpg.de/886719/MicelleFormation.pdf
2. What Is A Micelle in Biochemistry, David H. Nguyen, April 25, 2017:
http://sciencing.com/micelle-biochemistry-17102.html
3. University of California Museum of Paleontology, The Achaean Eon and the Hadean: http://www.ucmp.berkeley.edu/precambrian/archean_hadean.php
4. BBC: Nature and Prehistoric Life:
http://www.bbc.co.uk/nature/history_of_the_earth/Archean
5. Clay-armored bubbles may have formed first protocells: Minerals could have played a key role in the origins of life, Science Daily:
https://www.sciencedaily.com/releases/2011/02/110207073744.htm
6. NCBI: Transcription and RNApolymerase:
https://www.ncbi.nlm.nih.gov/books/NBK22085/
7. University of Colorado, Boulder,
http://www.colorado.edu/today/2005/04/05/cu-study-shows-early-earth-atmosphere-hydrogen-rich-favorable-life
8. Astrobiology Magazine, Dec. 2, 2011http://www.astrobio.net/geology/earths-early-atmosphere/
9. Universe Today, https://www.universetoday.com/26659/earths-early-atmosphere/
10. C.B. Thaxton, W.L. Bradley, and R.L. The Philosophical Library, 1984 *The Mystery od Life's Origin*
11. Growth of stable order in eukaryotes from environmental energy, Robert A. Gatenby, et al, US National Library of Medicine, Feb. 28, 2015
12. A Production of Amino Acids under Possible Primitive Earth Conditions, Stanley L. Miller, Science May 15, 1953http://www.abenteuer-universum.de/pdf/miller_1953.pdf
13. Following the Path of Discovery, Miller-Urey Experiment, Amino Acids and the Origins of Life on Earth,
http://www.julianrubin.com/bigten/miller_urey_experiment.html

14. NASA Technical Reports Server, Rosa Larralde, Stanley Miller, et al, Rates of Decomposition of Ribose and other Sugars: Implications for Chemical Evolution, Aug. 1, 1995, https://ntrs.nasa.gov/search.jsp?R=19980033941

Chapter 13

1. HHMI Researcher Jack Szostak Wind Nobel Prize in Physiology; Oct. 5, 2009 http://www.hhmi.org/news/hhmi-researcher-jack-szostak-wins-2009-nobel-prize-physiology-or-medicine
2. Rates of Decomposition of Ribose and other Sugars: Implications for Chemical Evolution: August 1, 1995, NASA Technical Reports Server https://ntrs.nasa.gov/search.jsp?R=19980033941
3. MRC Laboratory of Molecular Biology, John Sutherland awarded the Royal Society of Chemistry Tilden Prize 2011 http://www2.mrc-lmb.cam.ac.uk/john-sutherland-awarded-the-royal-society-of-chemistry-tilden-prize-2011/

Proof

Made in the USA
Columbia, SC
30 March 2018